S0-AZS-835

North America

midsize atlas

Key to Map Pages .... Inside front cover

How to Use this Atlas 2

Legend 3

Listing of State and City Maps....................................... 3

United States Interstate Map....................................... 4–5

Canada Highway Map.. 6–7

Maps.. 8–93

    United States 8–75

    BONUS MAPS! *Washington DC to Boston*
    Large-Scale Northeast Corridor Maps 68–71

    Canada... 75–91

    Mexico... 92–93

Index with City and Local Maps 93–125

    United States 94–122

    Canada... 123–125

    Mexico... 93

Distances and Driving Times across North America 126

City to City Distance Chart 127

Tire Tips 128

ROAD MAPS are organized geographically. *United States, Canada, and Mexico road maps are organized in a grid layout, starting in the northwest of each country. To find your way, use either the* **Key to Map Pages** *inside the front cover, the* **Listing of State and City Maps** *on page 3, or the* **index** *in the back of the atlas.*

COUNTRY COLORS
Colors represent countries throughout the atlas.
Red → Canada
Green → Mexico
Blue → United States
Purple → United States (Northeast Corridor)

MAP SCALES
Scale bars use consistent increments throughout the atlas for quick and easy scale comparison between regions.

DRIVING DISTANCES
Use this chart to check driving distances between major cities within each map. Refer to distance and driving time information at the back of the atlas for travel over greater distances.

LOCATOR MAPS
A quick glance at this miniature map lets you check which states and/or provinces are shown on each page.

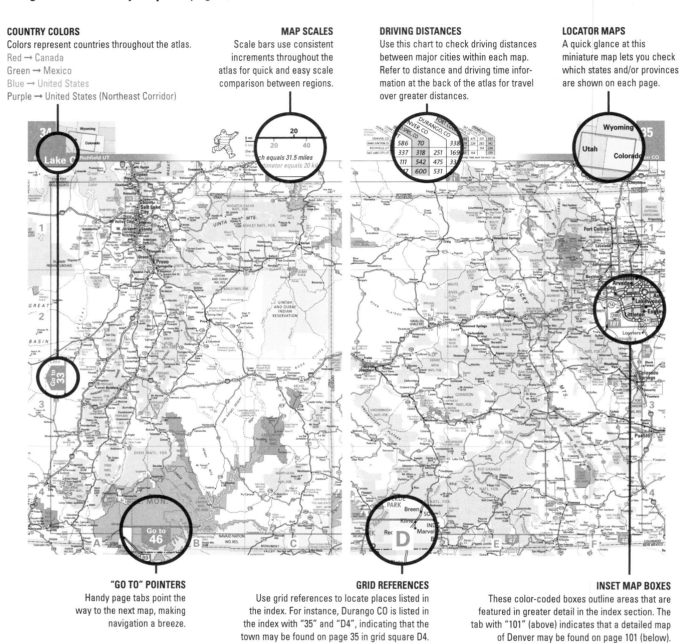

"GO TO" POINTERS
Handy page tabs point the way to the next map, making navigation a breeze.

GRID REFERENCES
Use grid references to locate places listed in the index. For instance, Durango CO is listed in the index with "35" and "D4", indicating that the town may be found on page 35 in grid square D4.

INSET MAP BOXES
These color-coded boxes outline areas that are featured in greater detail in the index section. The tab with "101" (above) indicates that a detailed map of Denver may be found on page 101 (below).

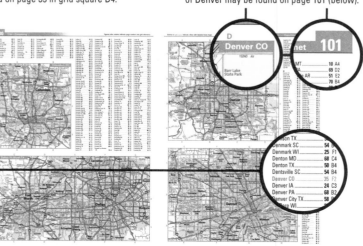

HOW THE INDEX WORKS
Cities and towns are listed alphabetically, with separate indexes for the United States, Canada, and Mexico. Figures after entries indicate page number and grid reference. Entries in bold color indicate cities with detailed inset maps.

TRANSPORTATION

HIGHWAYS

Freeway

Tollway; Toll Booth
City maps only

Under Construction

Interchange and Exit Number

Rest Area; Service Area
City maps only; Yellow with facilities

Primary; Secondary Highway

Multilane Divided Highway
Primary and secondary highways only

Other Paved Road; Unpaved Road
Check conditions locally

HIGHWAY MARKERS

Interstate Route

U.S. Route; State or Provincial Route

County or Other Route

Business Route

Trans-Canada Highway; Yellowhead Route

Canadian Provincial Autoroute

Mexican Federal Route

OTHER SYMBOLS

Distances along Major Highways
Miles in U.S.; kilometers in Canada and Mexico

Tunnel; Pass

Airport

Auto Ferry; Passenger Ferry

FEATURES OF INTEREST

National Park

National Forest; National Grassland

Other Large Park or Recreation Area

Small State Park with and without Camping

Public Campsite
City maps only

Trail

Point of Interest

Visitor Information Center
City maps only

Public Golf Course; Private Golf Course
City maps only; Selected professional tournament location

Hospital
City maps only

Ski Area

CITIES AND TOWNS

National Capital; State or Provincial Capital

Cities, Towns, and Populated Places
Type size indicates relative importance

Urban Area
State and province maps only

Large Incorporated Cities
City maps only

OTHER MAP FEATURES

Inset Map Boxes and Page Tabs
Tab indicates page where detailed inset map may be found

"Go to" Pointers
Page tabs to direct you to adjacent maps

International; State or Province Boundary

Time Zone Boundary

Mountain Peak; Elevation
Feet in U.S.; meters in Canada and Mexico

Perennial; Intermittent River

Perennial; Intermittent or Dry Water Body

Dam; Swamp

UNITED STATES   pages 8–75

ALABAMA (AL)............... 52–53, 64–65
  Birmingham 96
ALASKA (AK)...........................74–75
ARIZONA (AZ).............. 45–47, 56–57
  Phoenix...........................117
ARKANSAS (AR) 51–52, 63
  Little Rock........................105
CALIFORNIA (CA)... 17–18, 31–32, 44–45, 56–57
  Los Angeles 106–107
  Sacramento 119
  San Diego........................121
  San Francisco Bay Area 122
COLORADO (CO)............ 34–36, 47–48
  Denver...........................101
CONNECTICUT (CT)............ 29, 43, 70–71
  Hartford..........................102
DELAWARE (DE)................43, 68–69
DISTRICT OF COLUMBIA (DC)..........42, 68
  Washington 124
FLORIDA (FL)....................64–67
  Jacksonville......................103
  Miami/Fort Lauderdale.............109
  Orlando 115
  Tampa/St. Petersburg..............123
GEORGIA (GA) 53–54, 65–66
  Atlanta...........................94
HAWAII (HI) 72–73
  Honolulu 102
IDAHO (ID)................. 9–10, 18–20
ILLINOIS (IL) 25, 39–40, 52
  Chicago...........................98
INDIANA (IN) 25–26, 40
  Indianapolis 103
IOWA (IA) 23–25, 38–39
KANSAS (KS) 36–38, 49–51
  Kansas City.......................104
KENTUCKY (KY) 40–41, 52–54
  Louisville........................108
LOUISIANA (LA) 63–64
  New Orleans 111
MAINE (ME) 29–30
MARYLAND (MD) 42–43, 68
  Baltimore.........................95
  Washington DC 124
MASSACHUSETTS (MA) 29, 70–71
  Boston............................96
MICHIGAN (MI) 15–16, 26
  Detroit...........................101
MINNESOTA (MN) 13–15, 23–24
  Minneapolis/St. Paul 110
MISSISSIPPI (MS)............. 51–52, 63–64
MISSOURI (MO)............. 38–39, 51–52
  Kansas City.......................104
  St. Louis.........................120
MONTANA (MT)............. 9–12, 19–21
NEBRASKA (NE)............. 21–23, 36–38
NEVADA (NV)............. 18–19, 32–33, 45
  Las Vegas.........................105
NEW HAMPSHIRE (NH)..........29–30, 70–71
NEW JERSEY (NJ)............ 28, 43, 68–70
  Newark (New York NY) 112
NEW MEXICO (NM)............. 47–48, 56–58
  Albuquerque 94
NEW YORK (NY)........... 27–29, 43, 69–71
  Buffalo/Niagara Falls 97
  New York 112–113

NORTH CAROLINA (NC) 53–55
  Charlotte.........................97
NORTH DAKOTA (ND) 12–14
OHIO (OH) 26–27, 40–41
  Cincinnati 99
  Cleveland 99
  Columbus 100
OKLAHOMA (OK)....................48–51
  Oklahoma City 114
OREGON (OR) 8–9, 17–19
  Portland 118
PENNSYLVANIA (PA) 27–28, 41–43, 68–69
  Philadelphia......................116
  Pittsburgh........................117
RHODE ISLAND (RI)...............29, 71
  Providence........................118
SOUTH CAROLINA (SC).............53–55, 66
  Charleston........................97
SOUTH DAKOTA (SD) 12–14, 21–23
TENNESSEE (TN) 52–54
  Memphis..........................108
  Nashville.........................111
TEXAS (TX) 48–51, 57–63
  Austin 95
  Dallas/Fort Worth 100
  Houston...........................102
  San Antonio 121
UTAH (UT) 19–20, 33–34, 45–47
  Salt Lake City....................120
VERMONT (VT)......................29, 70
VIRGINIA (VA).............. 41–43, 53–55, 68
  Norfolk/Hampton Roads.............114
  Richmond 119
  Washington DC 124
WASHINGTON (WA)......................8–9
  Seattle/Tacoma 123
WEST VIRGINIA (WV)...............41–42
WISCONSIN (WI).............. 14–16, 24–25
  Milwaukee 109
WYOMING (WY) 20–22, 34–36
PUERTO RICO (PR).........................93

CANADA   pages 75–91

ALBERTA (AB) 75–76, 79–80
BRITISH COLUMBIA (BC)........ 75–76, 78–79
  Vancouver 125
MANITOBA (MB)....................77, 81–82
NEW BRUNSWICK (NB)................88–90
NEWFOUNDLAND & LABRADOR (NL)....89–91
NORTHWEST TERRITORIES (NT) 75–77
NOVA SCOTIA (NS)...................88–89
NUNAVUT (NU) 6–7, 75, 77
ONTARIO (ON) 77, 81–86
  Niagara Falls (Buffalo NY)..........97
  Toronto 125
  Windsor (Detroit MI) 101
PRINCE EDWARD ISLAND (PE).........88–90
QUÉBEC (QC) 83, 85–91
  Montréal..........................124
SASKATCHEWAN (SK) 76–77, 80–81
YUKON TERRITORY (YT).....................75

MEXICO   pages 92–93

México............................125

0 mi — 125 — 250 — 375
0 km — 125 — 250 — 375 — 500

One inch equals 204.6 miles
One centimeter equals 129.4 kilometers

0 150 300 mi
0 150 300 km

0 50 100 mi
0 50 100 km

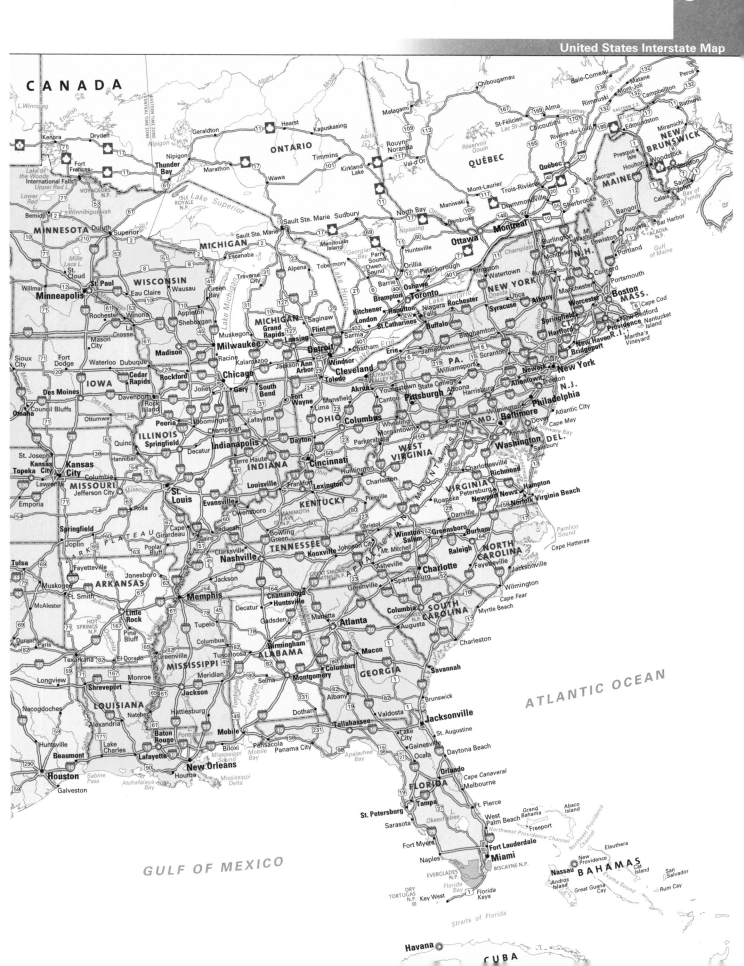

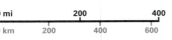

One inch equals 235.8 miles/Un pouce équivaut à 235.8 milles
One cm equals 149.3 km/Un cm équivaut à 149.3 km

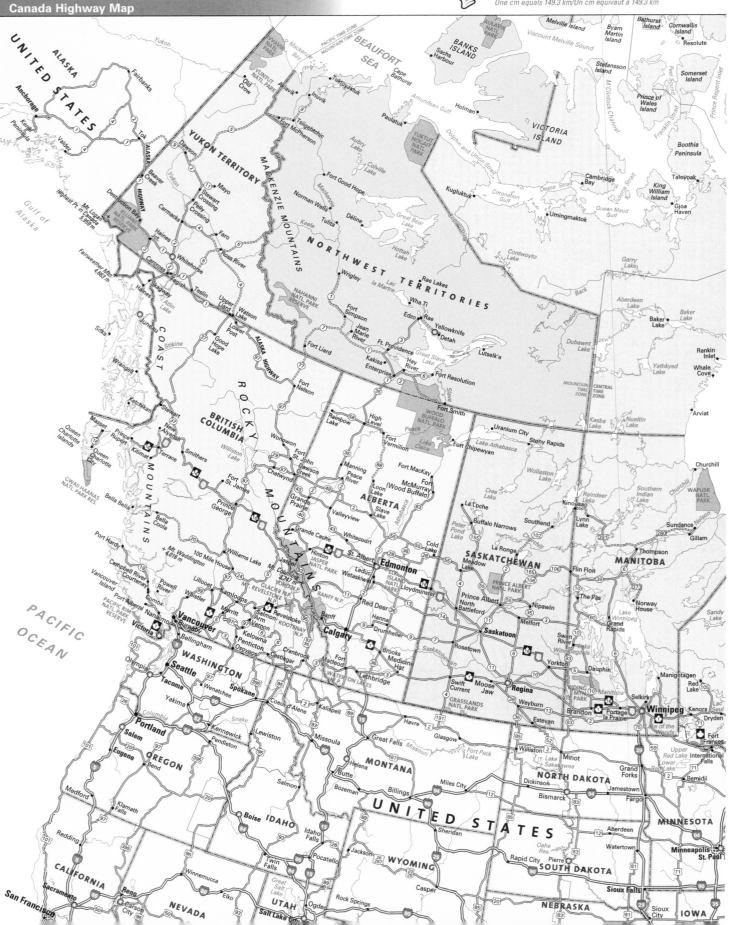

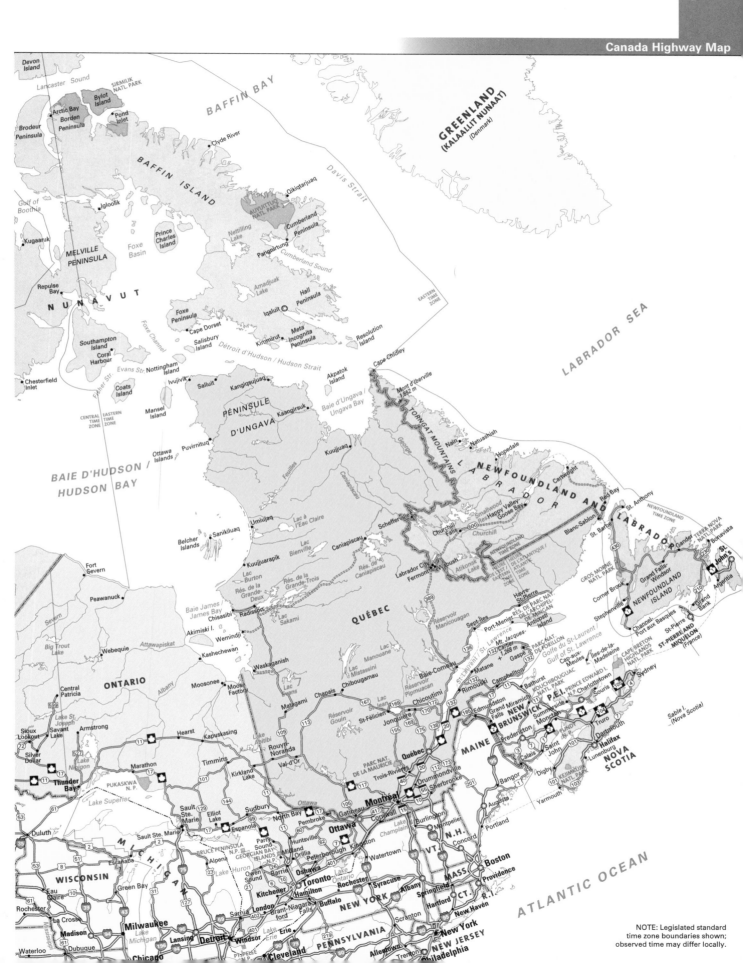

NOTE: Legislated standard
time zone boundaries shown;
observed time may differ locally.

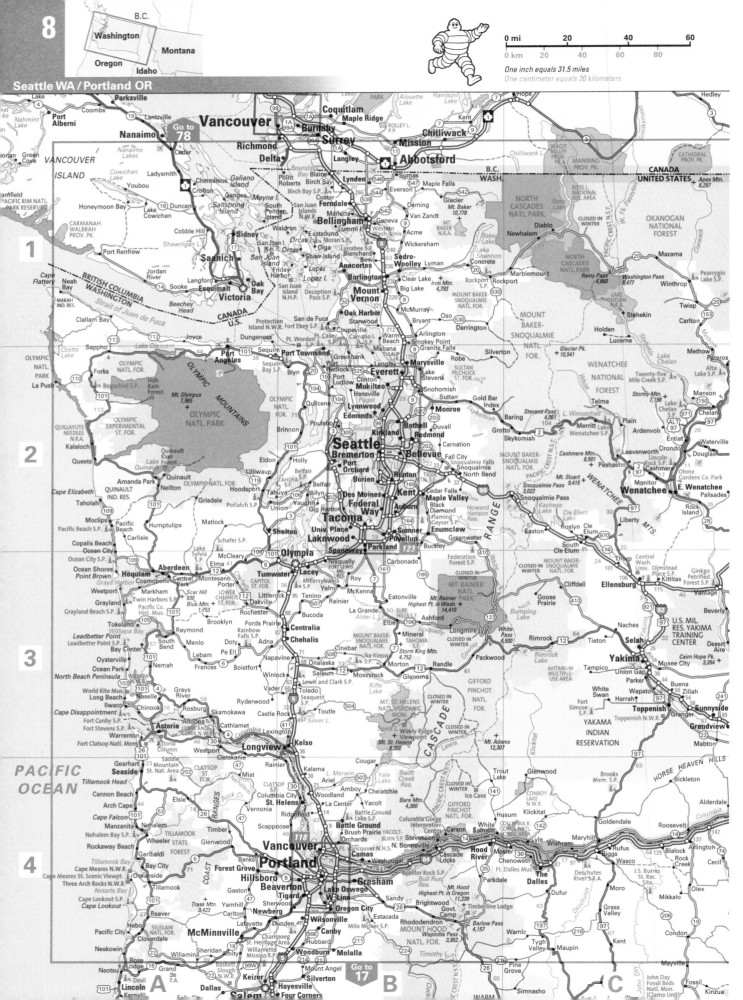

B.C.
Washington
Montana
Oregon
Idaho

0 mi   20   40   60
0 km   20   40   60   80
One inch equals 31.5 miles
One centimeter equals 20 kilometers

PACIFIC
OCEAN

VANCOUVER ISLAND

BRITISH COLUMBIA
WASHINGTON

CANADA
U.S.

Go to 78

Vancouver
Nanaimo
Richmond
Delta
Surrey
Burnaby
Maple Ridge
Coquitlam
Mission
Chilliwack
Abbotsford
Langley
Hope

Parksville
Coombs
Port Alberni
Lantzville
Nahmint Lake
Green Cove

Honeymoon Bay
Youbou
Ladysmith
Chemainus
Crofton
Cedar

Duncan
Cobble Hill
Sidney
Saanich
Langford
Esquimalt
Victoria
Oak Bay

Cape Flattery
Neah Bay
Clallam Bay
Sappho
Forks
La Push
Joyce
Port Angeles
Sequim
Port Townsend

OLYMPIC
NATL. FOR.
OLYMPIC
NATL. PARK
OLYMPIC MOUNTAINS
Mt. Olympus 7,965

Bellingham
Ferndale
Lynden
Sumas
Blaine
Birch Bay
Deming
Glacier
Mt. Baker 10,778

NORTH CASCADES NATL. PARK

OKANOGAN NATIONAL FOREST

Mount Vernon
Burlington
Anacortes
Sedro-Woolley
Oak Harbor
Coupeville
Stanwood
Arlington
Darrington

Mazama
Winthrop
Twisp
Carlton
Stehekin
Holden
Lucerne

Everett
Marysville
Mukilteo
Lynnwood
Edmonds
Monroe
Snohomish
Sultan
Gold Bar
Index
Skykomish

WENATCHEE
NATIONAL
FOREST

Leavenworth
Peshastin
Cashmere
Wenatchee
E. Wenatchee

Seattle
Bremerton
Bellevue
Kirkland
Redmond
Bothell
Renton
Kent
Burien
Des Moines
Federal Way
Tacoma
Auburn
Maple Valley
Enumclaw
Snoqualmie
North Bend

Snoqualmie Pass
Cle Elum
Roslyn
Easton
Ellensburg
Vantage

MOUNT BAKER-
SNOQUALMIE
NATL. FOR.

CASCADE RANGE

Olympia
Lacey
Tumwater
Shelton
Lakewood
Parkland
Spanaway
Puyallup
Sumner
Buckley

Aberdeen
Hoquiam
Cosmopolis
Montesano
Elma
McCleary

Ocean Shores
Westport
Grayland
Tokeland
Raymond
South Bend

Centralia
Chehalis
Napavine
Winlock
Toledo
Castle Rock

MT. RAINIER
NATL. PARK
Mt. Rainier Highest Pt. in Wash. 14,410

Yakima
Selah
Union Gap
Parker
Wapato
Toppenish
Granger
Grandview
Sunnyside
Zillah
Moxee City

YAKAMA
INDIAN
RESERVATION

U.S. MIL RES. YAKIMA
TRAINING CENTER

Long Beach
Ilwaco
Astoria
Warrenton
Seaside
Cannon Beach
Manzanita
Rockaway Beach
Garibaldi
Tillamook

Longview
Kelso
Kalama
Woodland
St. Helens
Scappoose

MT. ST. HELENS
NATL. VOLCANIC MON.

GIFFORD
PINCHOT
NATL.
FOR.

Mt. Adams 12,307

HORSE HEAVEN HILLS

Glenwood
Klickitat
Goldendale
Roosevelt
Alderdale

Portland
Vancouver
Gresham
Beaverton
Hillsboro
Forest Grove
Tigard
Lake Oswego
Oregon City
Wilsonville
Newberg
McMinnville
Canby
Woodburn
Molalla
Silverton

Camas
Washougal
Hood River
The Dalles
Stevenson
White Salmon
Carson
Bingen
Lyle
Mosier
Dufur
Maupin

Go to 17

Mt. Hood Highest Pt. in Oregon 11,239

MOUNT HOOD
NATL. FOR.

Lincoln City
Salem
Keizer
Dallas
Four Corners

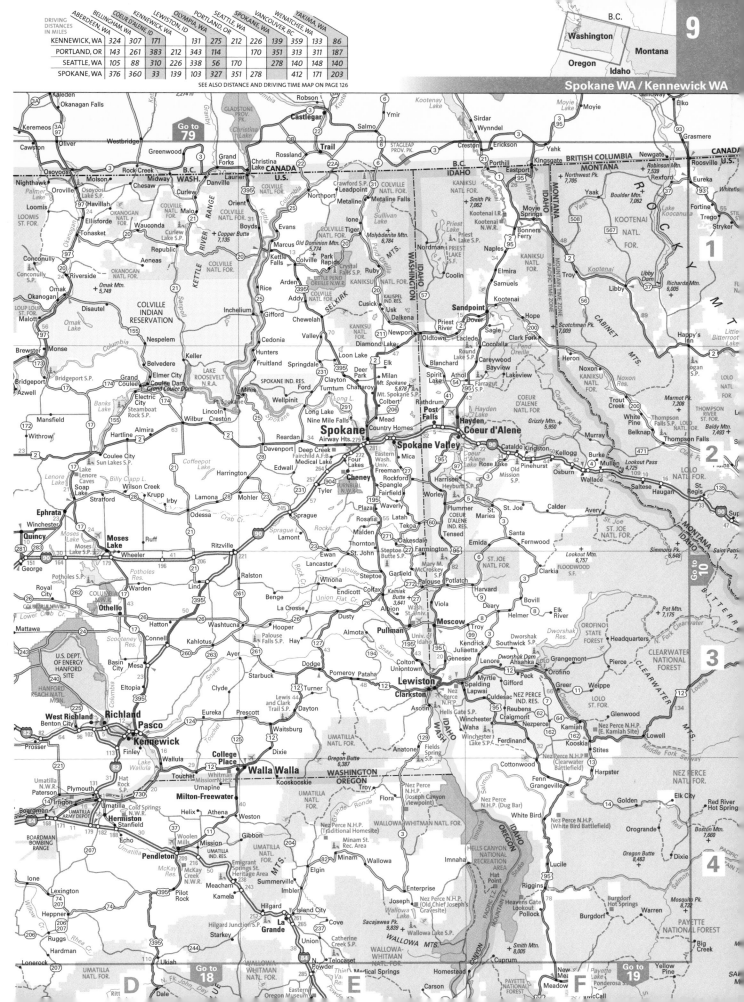

SEE ALSO DISTANCE AND DRIVING TIME MAP ON PAGE 126

DRIVING DISTANCES IN MILES	ABERDEEN, WA	BELLINGHAM, WA	COEUR D'ALENE, ID	KENNEWICK, WA	LEWISTON, ID	OLYMPIA, WA	PORTLAND, OR	SEATTLE, WA	SPOKANE, WA	VANCOUVER, BC	WENATCHEE, WA	YAKIMA, WA
KENNEWICK, WA	324	307	171		131	275	212	226	139	359	133	86
PORTLAND, OR	143	261	383	212	343	114		170	351	313	311	187
SEATTLE, WA	105	88	310	226	338	56	170		278	140	148	140
SPOKANE, WA	376	360	33	139	103	328	351	278		412	171	203

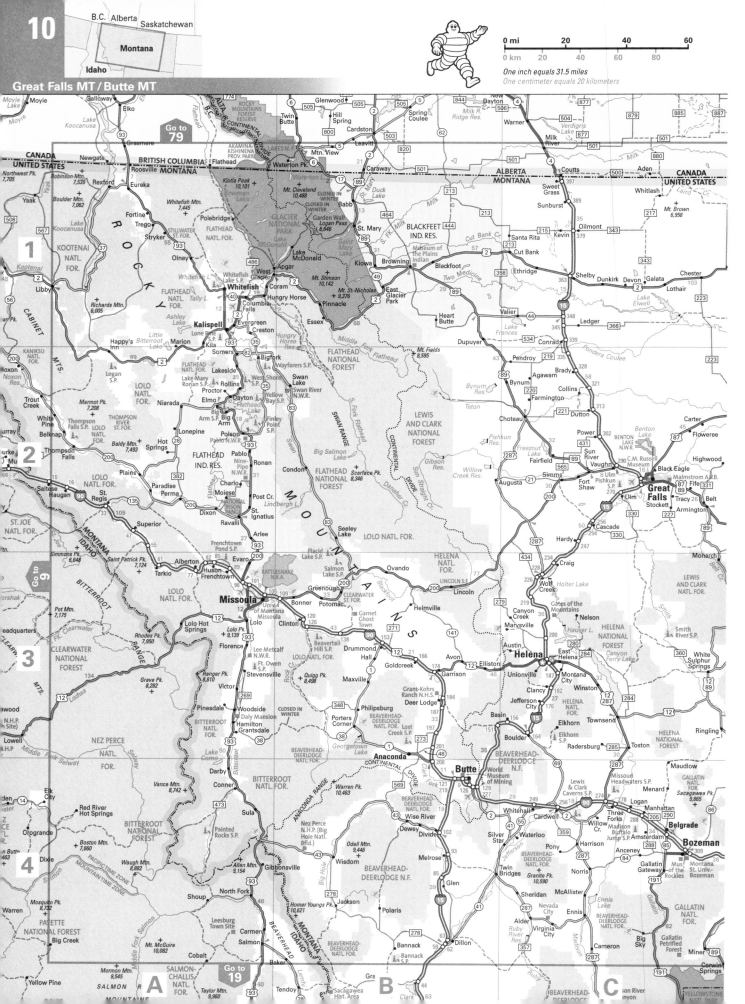

DRIVING DISTANCES IN MILES	BILLINGS, MT	BOZEMAN, MT	BUTTE, MT	GLASGOW, MT	GREAT FALLS, MT	HAVRE, MT	HELENA, MT	KALISPELL, MT	LEWISTOWN, MT	MISSOULA, MT	SALMON, ID	SHELBY, MT
BILLINGS, MT		141	223	277	222	249	235	454	125	340	372	304
BUTTE, MT	223	81		430	153	271	71	232	247	118	150	235
GLASGOW, MT	277	368	430		277	159	362	418	203	476	579	261
KALISPELL, MT	454	312	232	418	222	259	195		331	116	255	157

SEE ALSO DISTANCE AND DRIVING TIME MAP ON PAGE 126

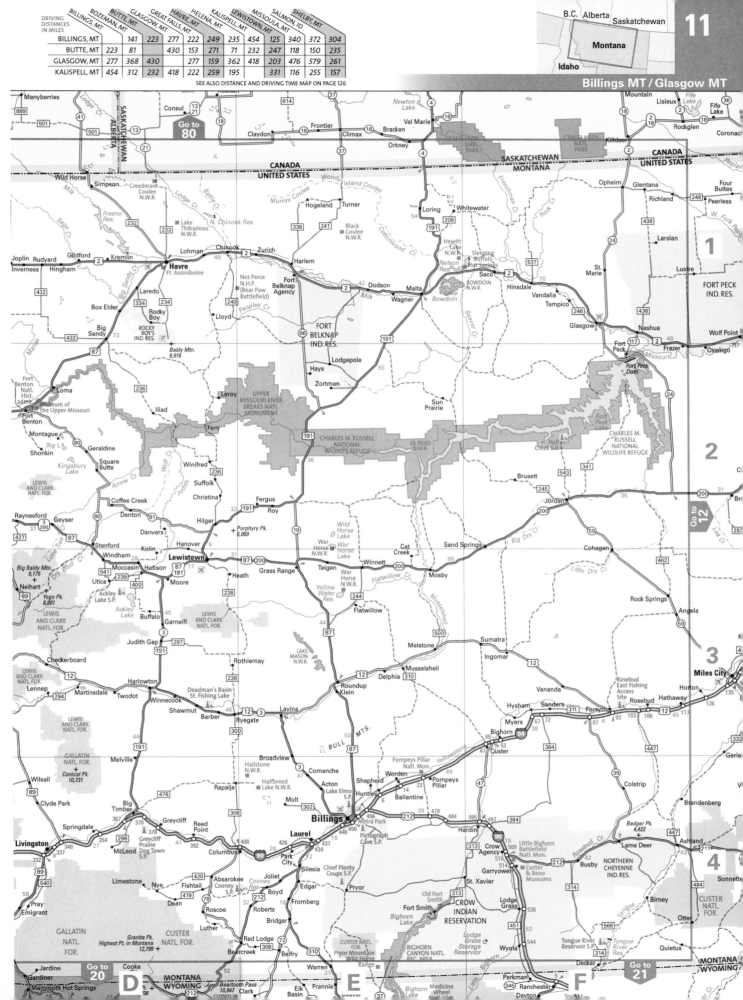

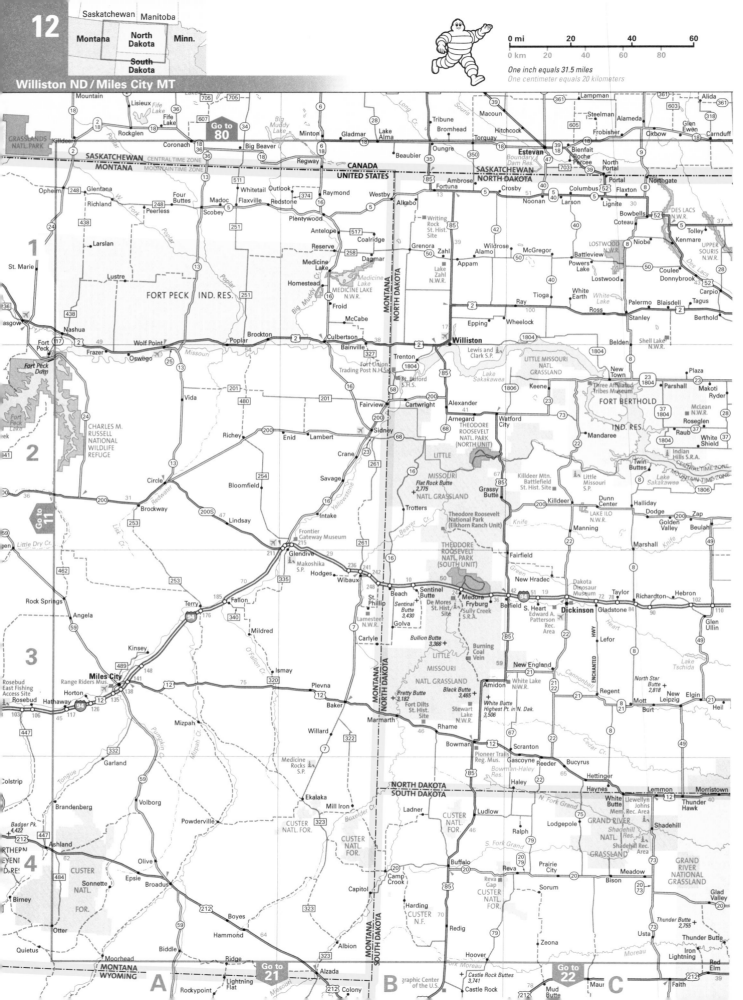

0 mi 20 40 60

0 km 20 40 60 80

One inch equals 31.5 miles
One centimeter equals 20 kilometers

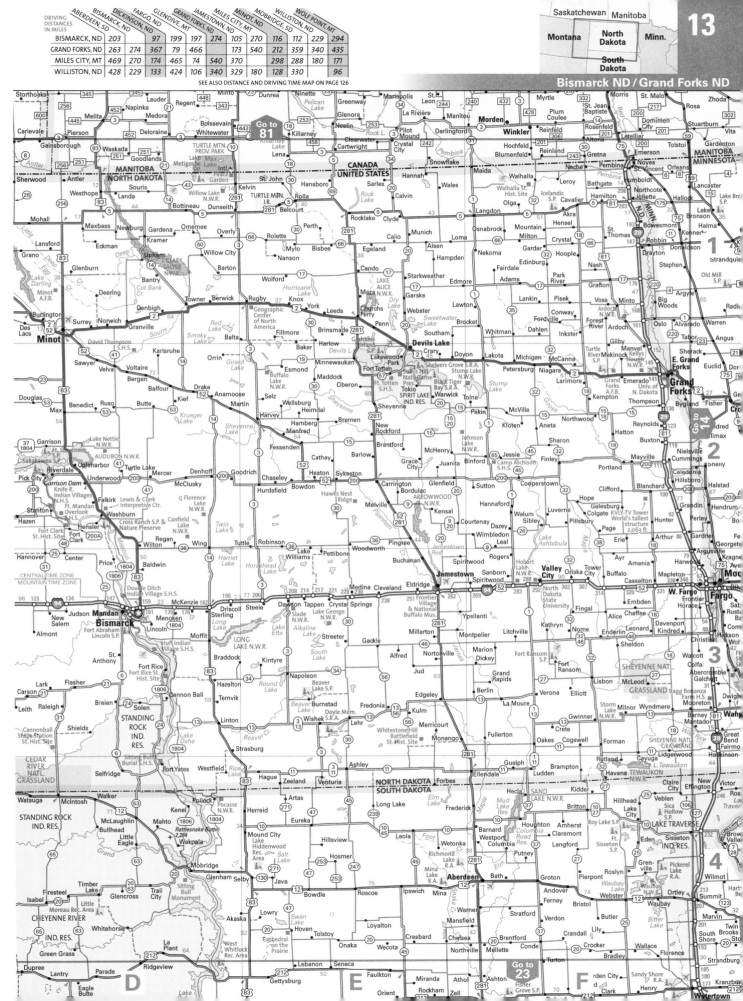

DRIVING DISTANCES IN MILES

	ABERDEEN, SD	BISMARCK, ND	DICKINSON, ND	FARGO, ND	GLENDIVE, MT	GRAND FORKS, ND	JAMESTOWN, ND	MILES CITY, MT	MINOT, ND	MOBRIDGE, SD	WILLISTON, ND	WOLF POINT, MT
BISMARCK, ND	203		97	199	197	274	105	270	116	112	229	294
GRAND FORKS, ND	263	274	367	79	466		173	540	212	359	340	435
MILES CITY, MT	469	270	174	465	74	540	370		298	288	180	171
WILLISTON, ND	428	229	133	424	106	340	329	180	128	330		96

SEE ALSO DISTANCE AND DRIVING TIME MAP ON PAGE 126

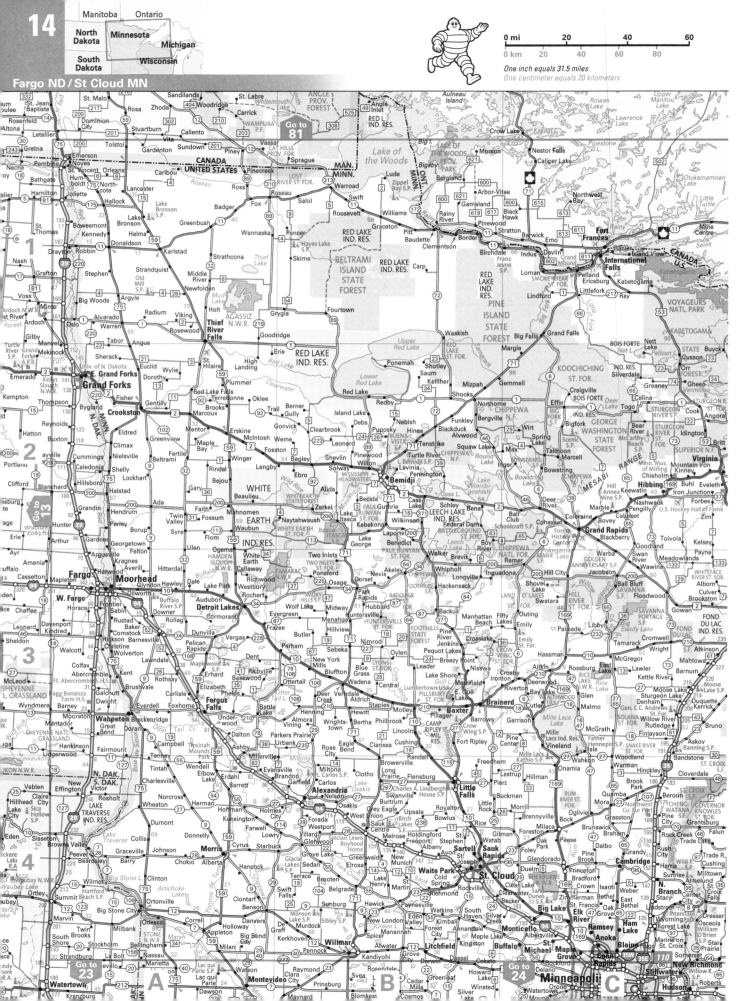

Manitoba Ontario
North Dakota Minnesota Michigan
South Dakota Wisconsin

0 mi 20 40 60
0 km 20 40 60 80
One inch equals 31.5 miles
One centimeter equals 20 kilometers

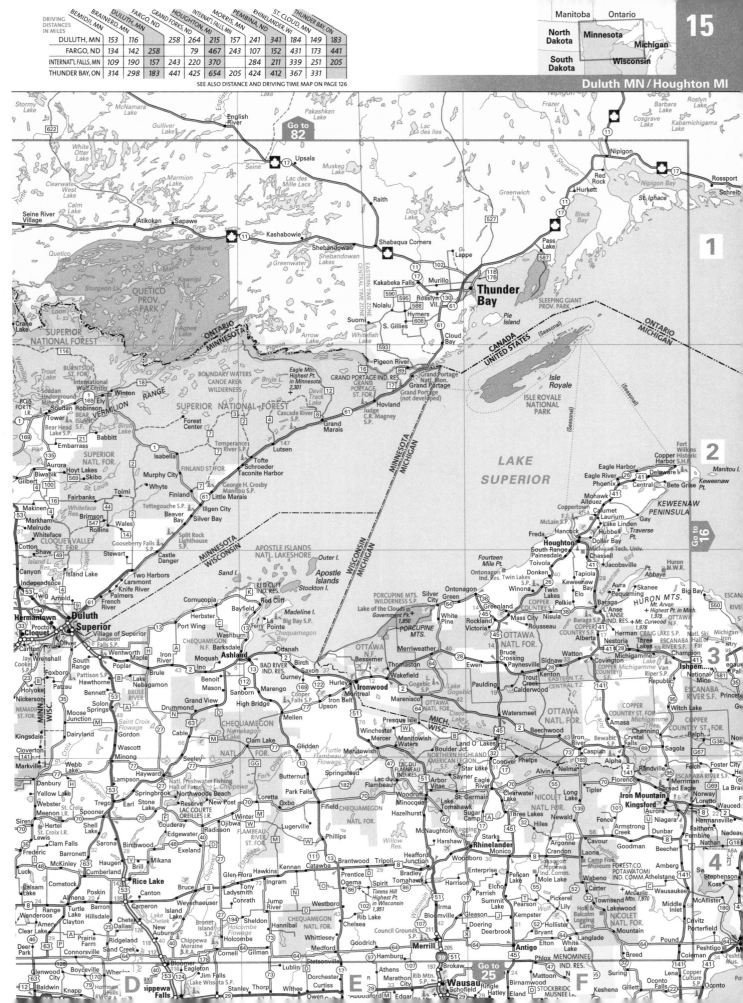

DRIVING DISTANCES IN MILES	BEMIDJI, MN	BRAINERD, MN	DULUTH, MN	FARGO, ND	GRAND FORKS, ND	HOUGHTON, MI	INTERNAT'L FALLS, MN	MORRIS, MN	PEMBINA, ND	RHINELANDER, WI	ST. CLOUD, MN	THUNDER BAY, ON
DULUTH, MN	153	116		258	264	215	157	241	341	184	149	183
FARGO, ND	134	142	258		79	467	243	107	152	431	173	441
INTERNAT'L FALLS, MN	109	190	157	243	220	370		284	211	339	251	205
THUNDER BAY, ON	314	298	183	441	425	654	205	424	412	367	331	

SEE ALSO DISTANCE AND DRIVING TIME MAP ON PAGE 126

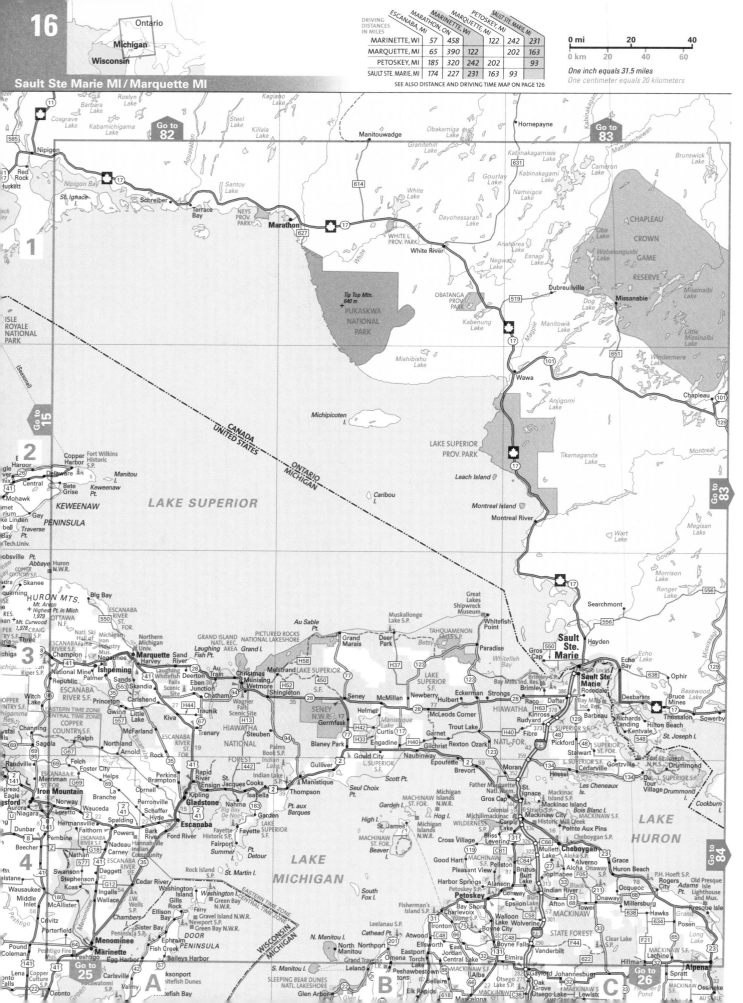

DRIVING DISTANCES IN MILES	ESCANABA, MI	MARATHON, ON	MARINETTE, WI	MARQUETTE, MI	PETOSKEY, MI	SAULT STE. MARIE, MI
MARINETTE, WI	57	458		122	242	231
MARQUETTE, MI	65	390	122		202	163
PETOSKEY, MI	185	320	242	202		93
SAULT STE. MARIE, MI	174	227	231	163	93	

SEE ALSO DISTANCE AND DRIVING TIME MAP ON PAGE 126

0 mi   20   40
0 km   20   40   60

One inch equals 31.5 miles
One centimeter equals 20 kilometers

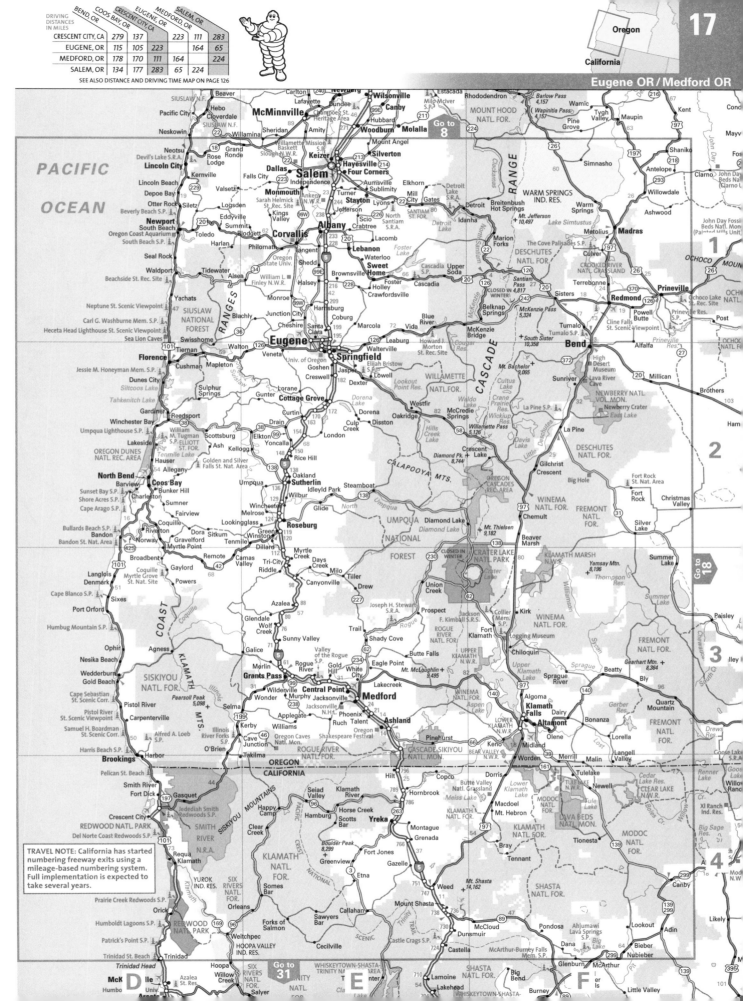

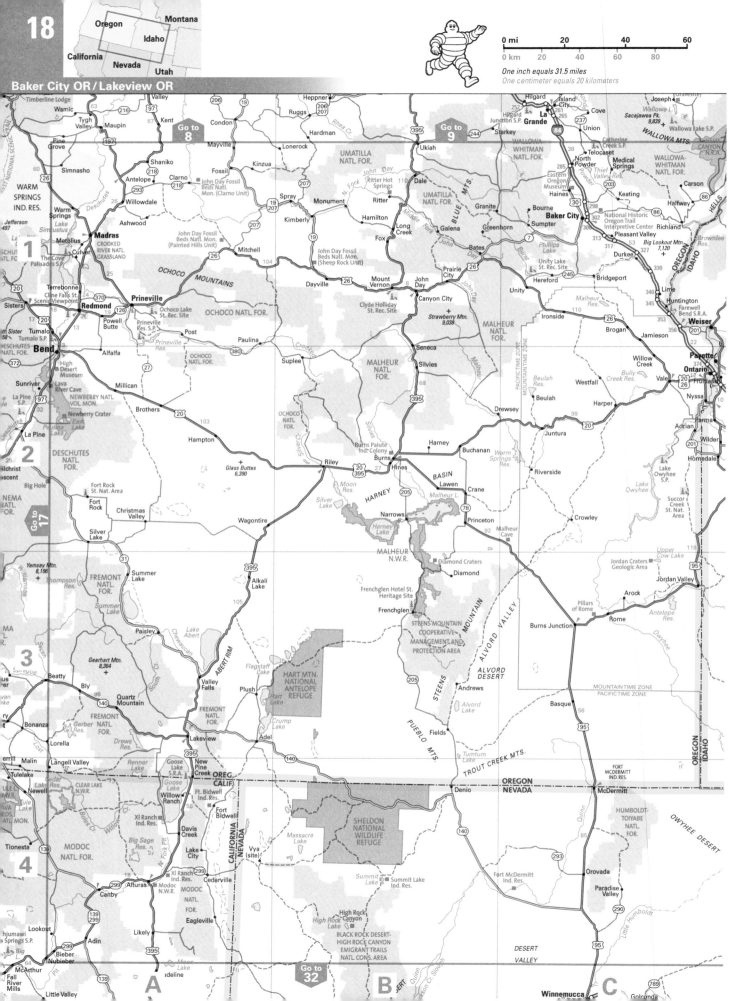

DRIVING DISTANCES IN MILES	ALTURAS, CA	BAKER CITY, OR	BEND, OR	BOISE, ID	BURNS, OR	LAKEVIEW, OR	ONTARIO, OR	POCATELLO, ID	SALMON, ID	SUN VALLEY, ID	TWIN FALLS, ID	WELLS, NV
BAKER CITY, OR	337		228	126	141	281	70	364	373	286	257	374
BEND, OR	233	228		330	142	177	272	568	576	490	460	550
BOISE, ID	384	126	330		188	328	58	241	247	163	134	251
POCATELLO, ID	621	364	568	241	426	566	295		217	190	116	233

SEE ALSO DISTANCE AND DRIVING TIME MAP ON PAGE 126

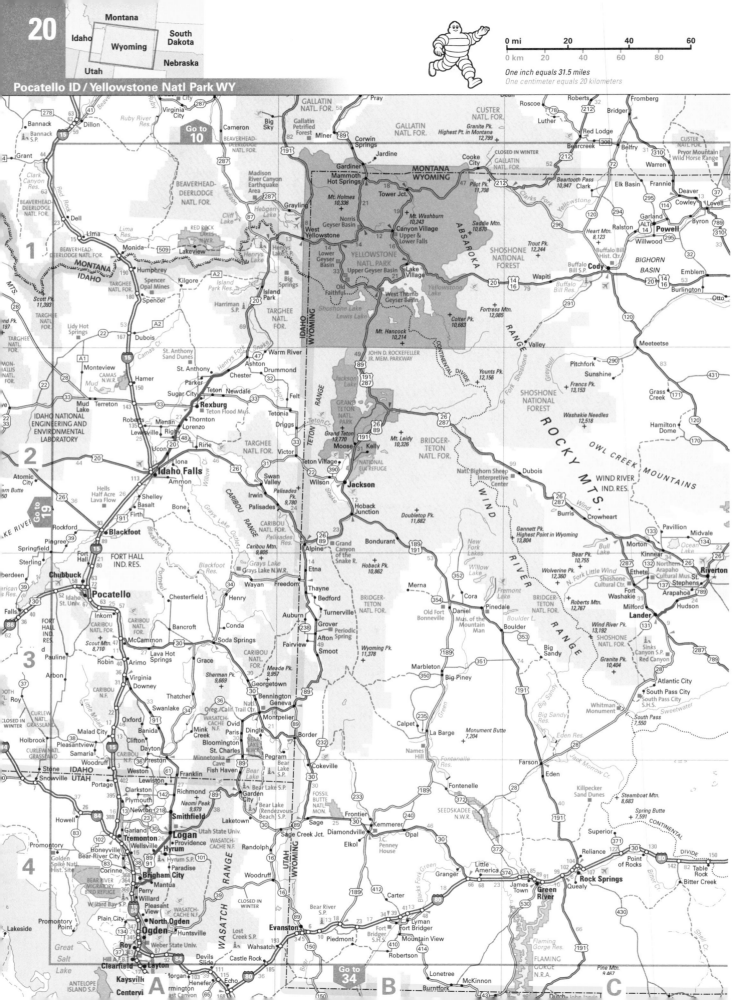

20

Montana
Idaho
Wyoming
South Dakota
Utah
Nebraska

Pocatello ID / Yellowstone Natl Park WY

0 mi   20   40   60
0 km   20   40   60   80
One inch equals 31.5 miles
One centimeter equals 20 kilometers

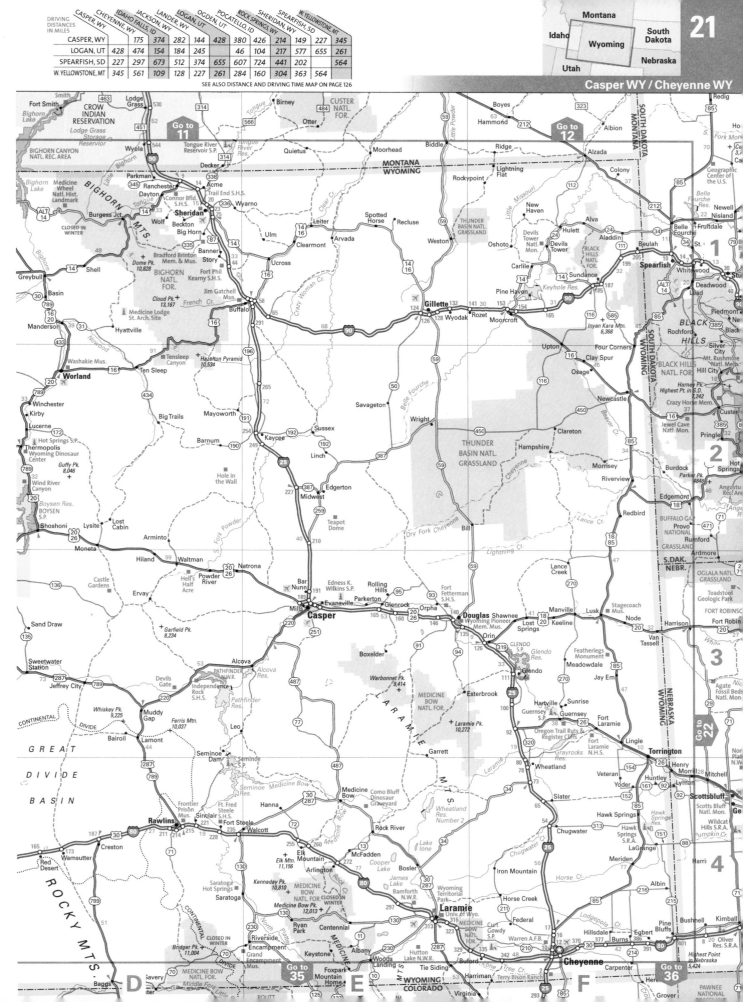

DRIVING DISTANCES IN MILES	CASPER, WY	CHEYENNE, WY	IDAHO FALLS, ID	JACKSON, WY	LANDER, WY	LOGAN, UT	OGDEN, UT	POCATELLO, ID	ROCK SPRINGS, WY	SHERIDAN, WY	SPEARFISH, SD	W. YELLOWSTONE, MT
CASPER, WY		175	374	282	144	428	380	426	214	149	227	345
LOGAN, UT	428	474	154	184	245		46	104	217	577	655	261
SPEARFISH, SD	227	297	673	512	374	655	607	724	441	202		564
W. YELLOWSTONE, MT	345	561	109	128	227	261	284	160	304	363	564	

SEE ALSO DISTANCE AND DRIVING TIME MAP ON PAGE 126

0 mi | 20 | 40 | 60
0 km | 20 | 40 | 60 | 80
One inch equals 31.5 miles
One centimeter equals 20 kilometers

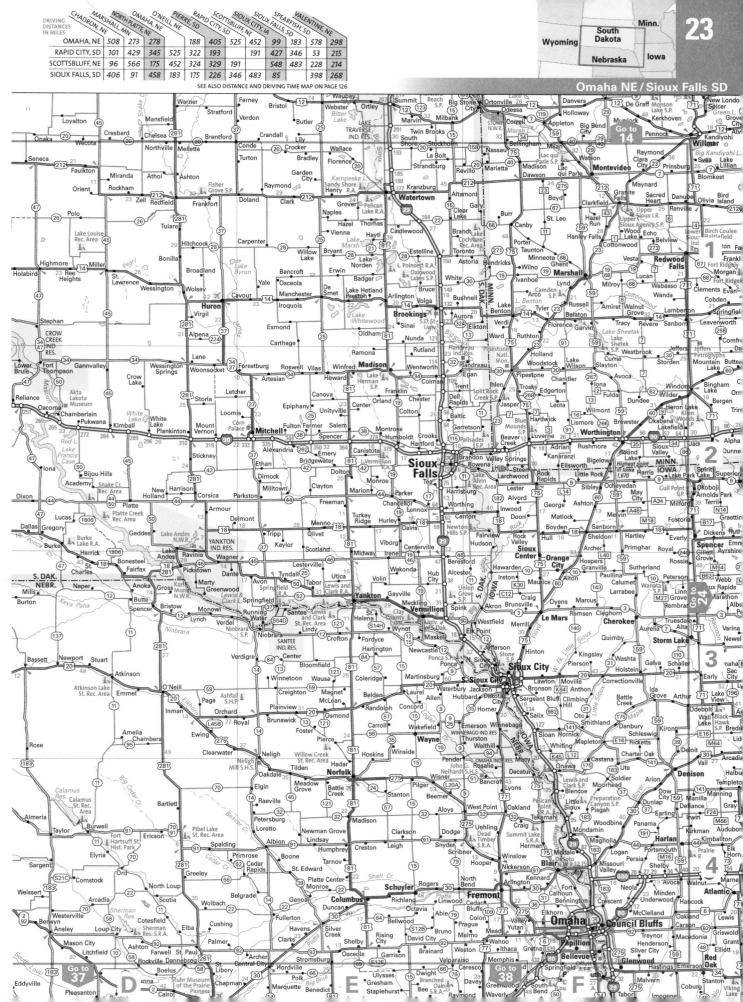

DRIVING DISTANCES IN MILES	CHADRON, NE	MARSHALL, MN	NORTH PLATTE, NE	OMAHA, NE	O'NEILL, NE	PIERRE, SD	RAPID CITY, SD	SCOTTSBLUFF, NE	SIOUX CITY, IA	SIOUX FALLS, SD	SPEARFISH, SD	VALENTINE, NE
OMAHA, NE	508	273	278		188	405	525	452	99	183	578	298
RAPID CITY, SD	101	429	345	525	322	193		191	427	346	53	215
SCOTTSBLUFF, NE	96	566	175	452	324	329	191		548	483	228	214
SIOUX FALLS, SD	406	91	458	183	175	226	346	483	85		398	268

SEE ALSO DISTANCE AND DRIVING TIME MAP ON PAGE 126

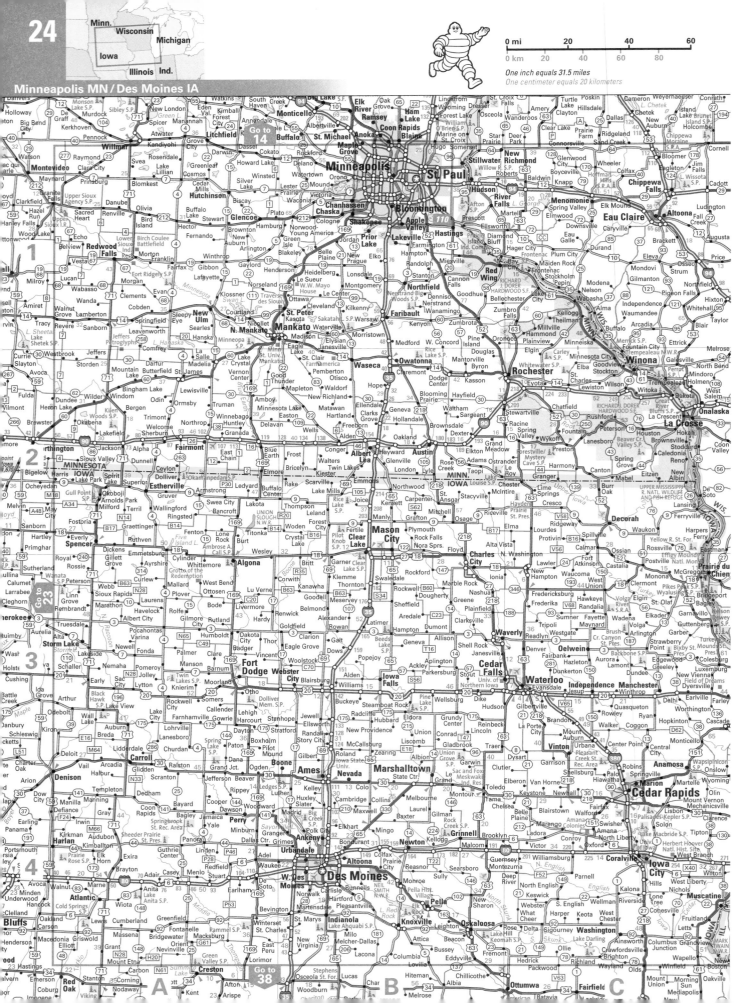

Minn.
Wisconsin
Michigan
Iowa
Illinois   Ind.

0 mi   20   40   60
0 km   20   40   60   80
One inch equals 31.5 miles
One centimeter equals 20 kilometers

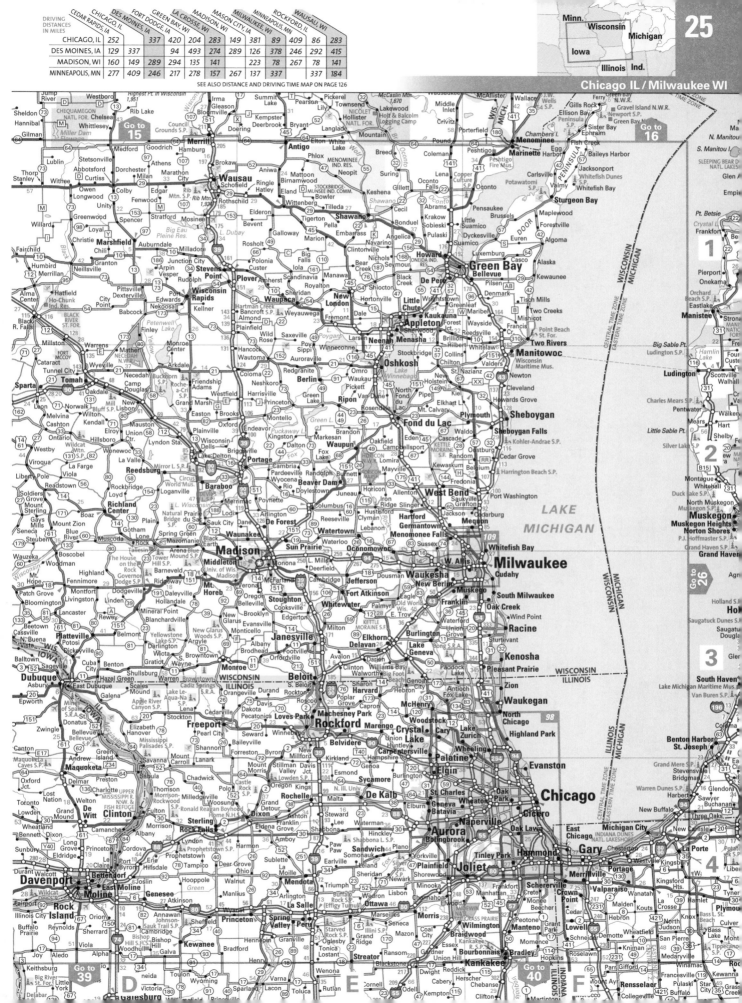

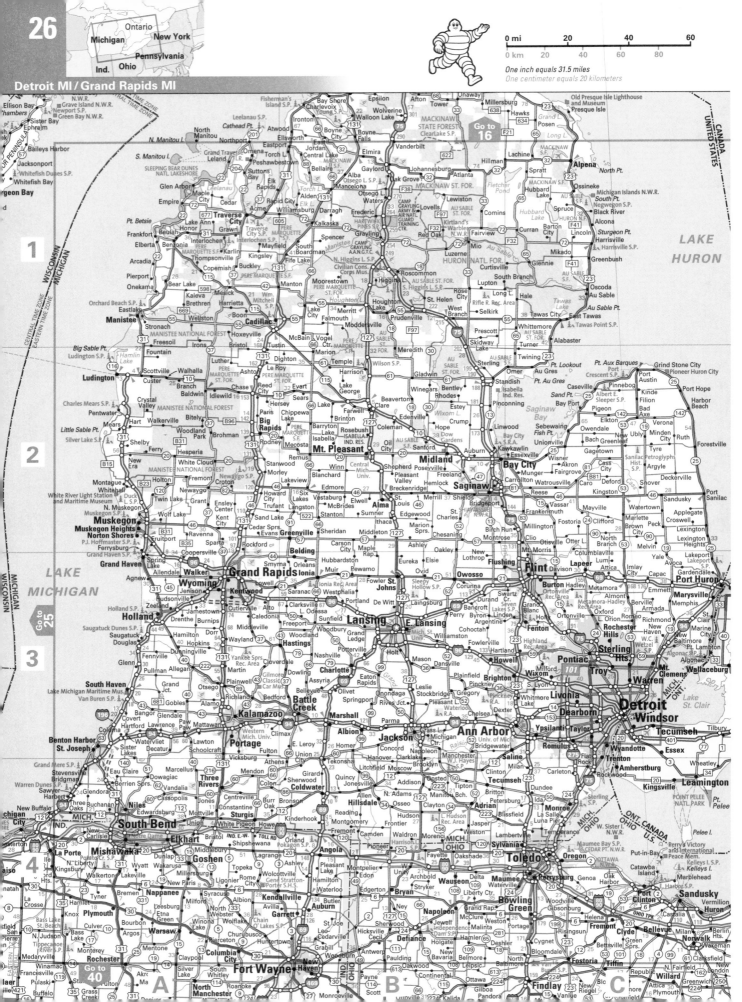

Ontario
Michigan   New York
Pennsylvania
Ind. Ohio

0 mi   20   40   60
0 km   20   40   60   80
One inch equals 31.5 miles
One centimeter equals 20 kilometers

Go to 16

Go to 25

Go to 40

LAKE HURON

LAKE MICHIGAN

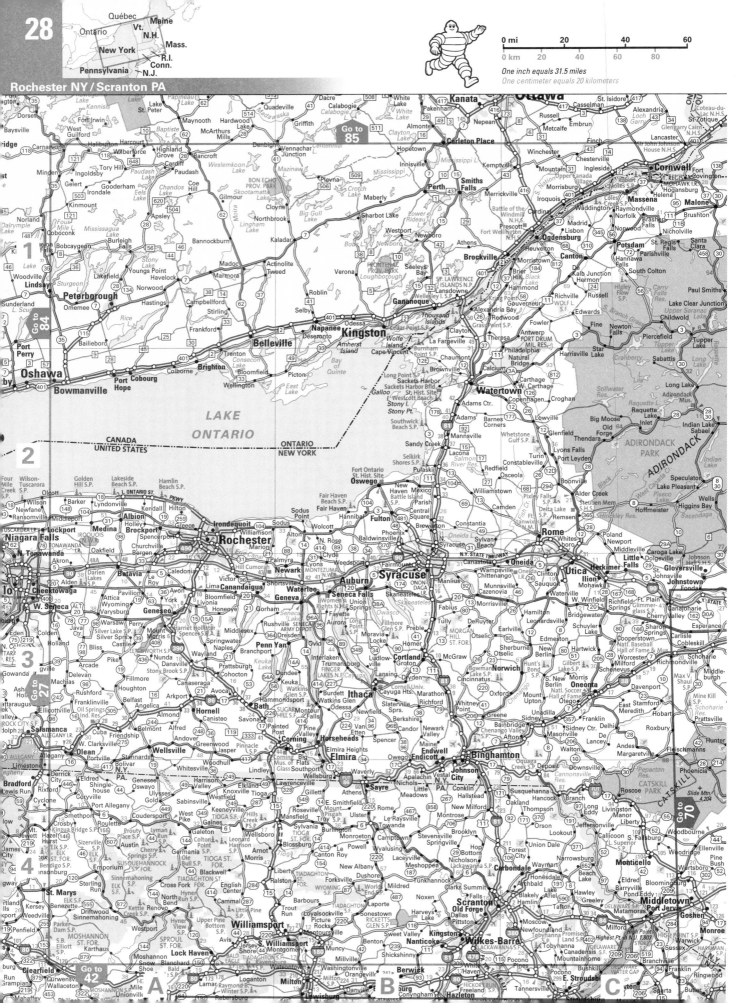

DRIVING DISTANCES IN MILES	ALBANY, NY	BOSTON, MA	BURLINGTON, VT	CORNWALL, ON	HARTFORD, CT	MANCHESTER, NH	PORTLAND, ME	PROVIDENCE, RI	ROCHESTER, NY	SCRANTON, PA	SYRACUSE, NY	WATERTOWN, NY
ALBANY, NY		172	147	227	111	145	270	170	228	180	146	179
BOSTON, MA	172		214	370	102	54	107	52	398	296	316	349
ROCHESTER, NY	228	398	311	264	337	364	496	396		220	88	146
SCRANTON, PA	180	296	328	316	194	326	394	272	220		135	198

SEE ALSO DISTANCE AND DRIVING TIME MAP ON PAGE 126

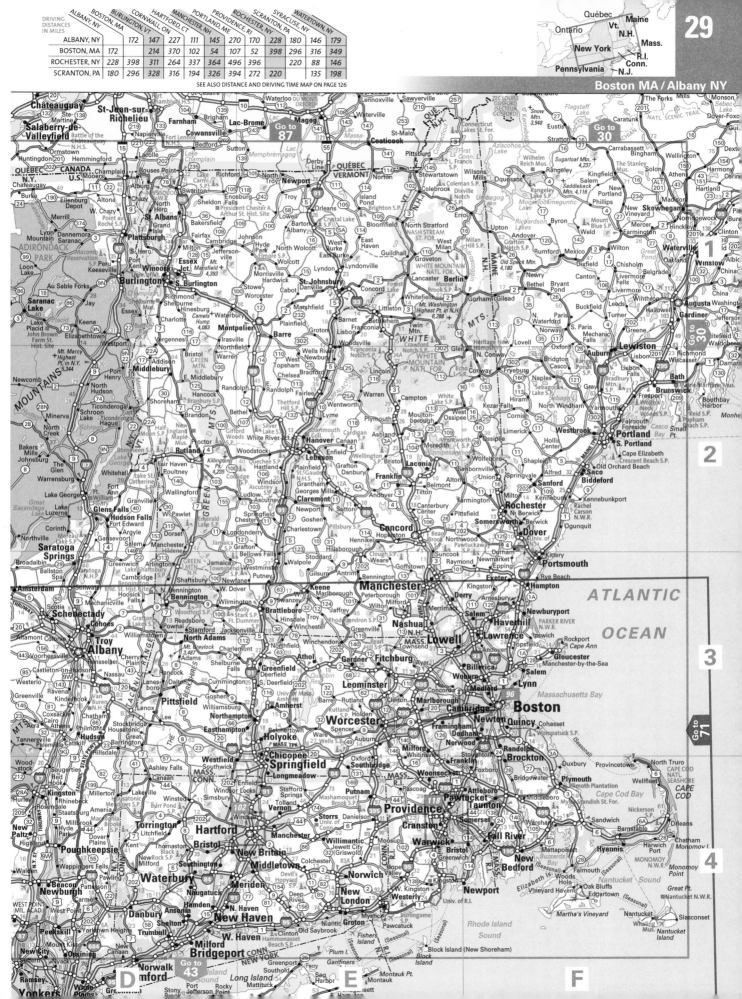

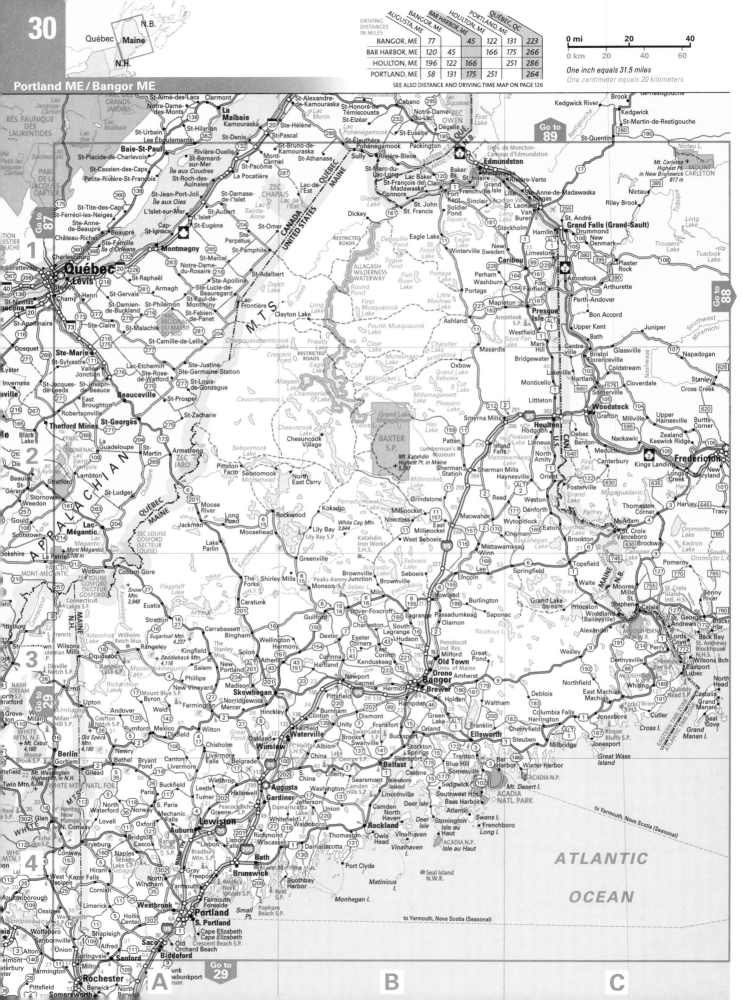

DRIVING DISTANCES IN MILES	EUREKA, CA	MONTEREY, CA	REDDING, CA	SACRAMENTO, CA	SAN FRANCISCO, CA	SAN JOSE, CA
EUREKA, CA		380	133	278	263	306
MONTEREY, CA	380		323	188	114	74
SACRAMENTO, CA	278	188	166		87	115
SAN FRANCISCO, CA	263	114	222	87		43

SEE ALSO DISTANCE AND DRIVING TIME MAP ON PAGE 126

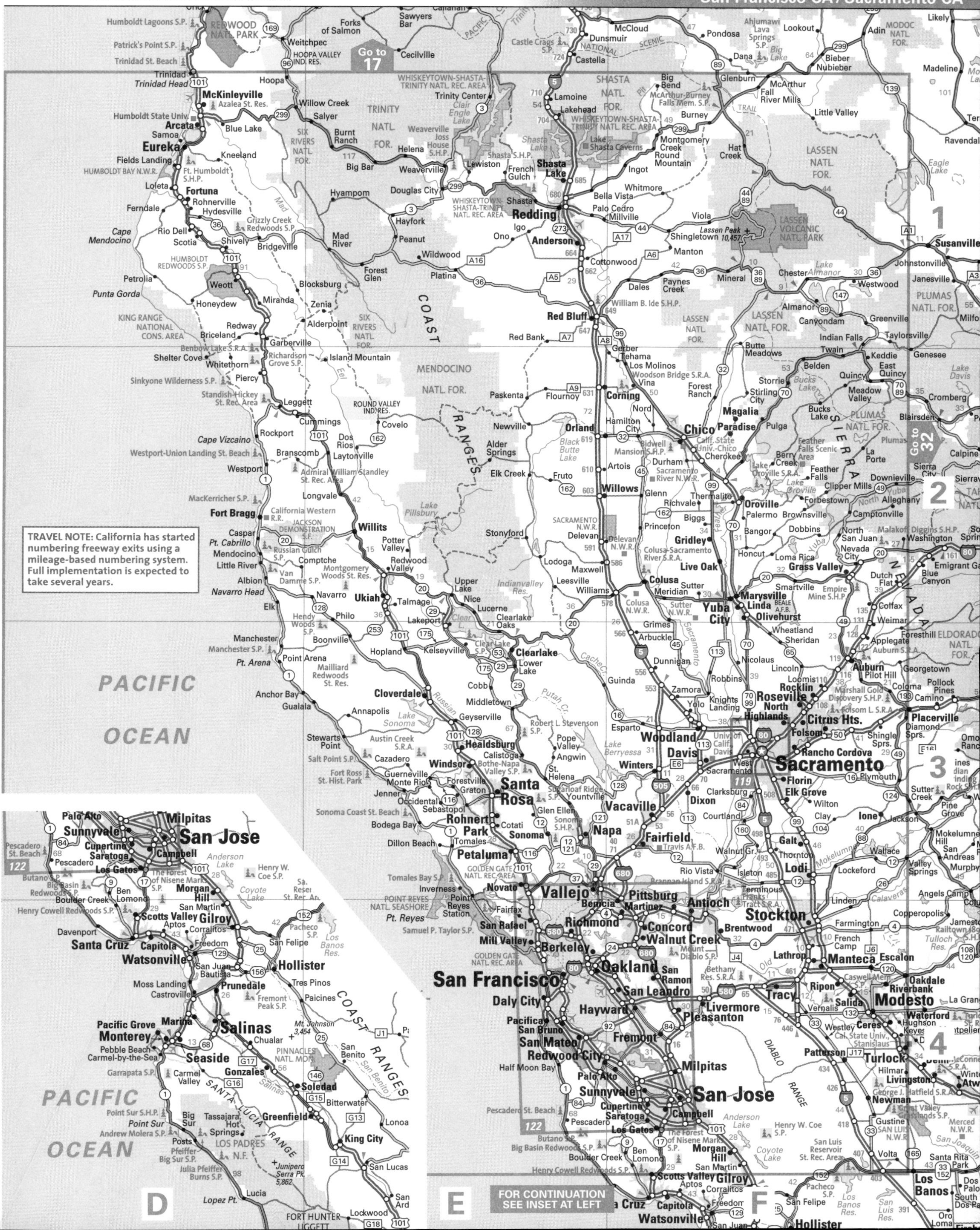

Go to 17

Go to 32

TRAVEL NOTE: California has started numbering freeway exits using a mileage-based numbering system. Full implementation is expected to take several years.

FOR CONTINUATION SEE INSET AT LEFT

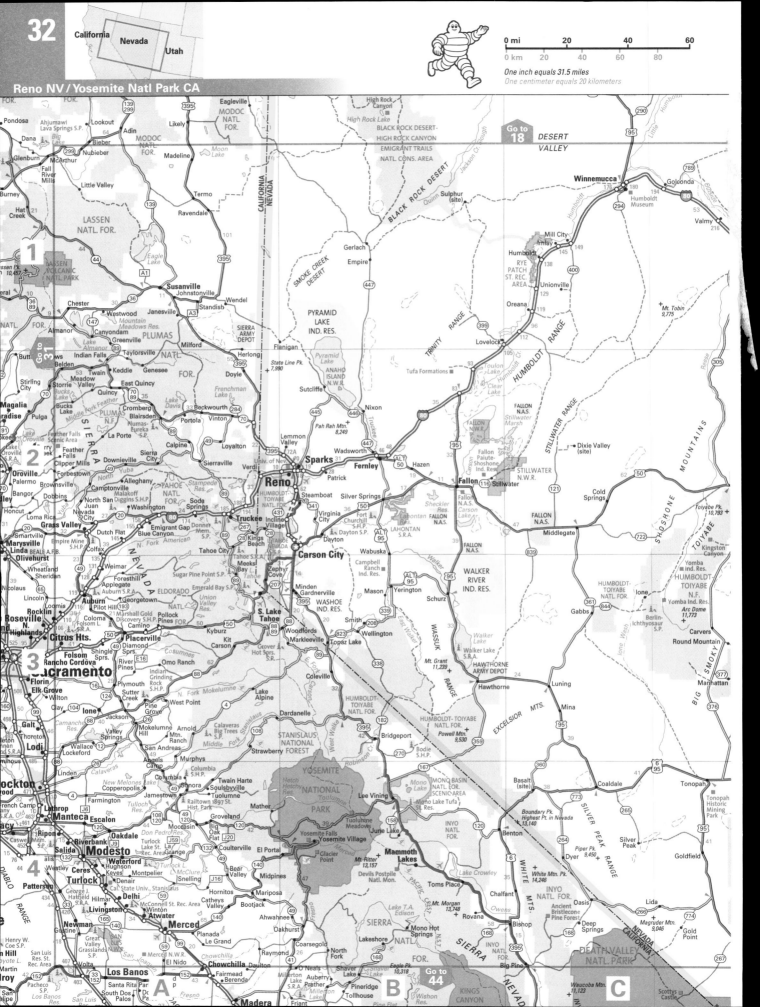

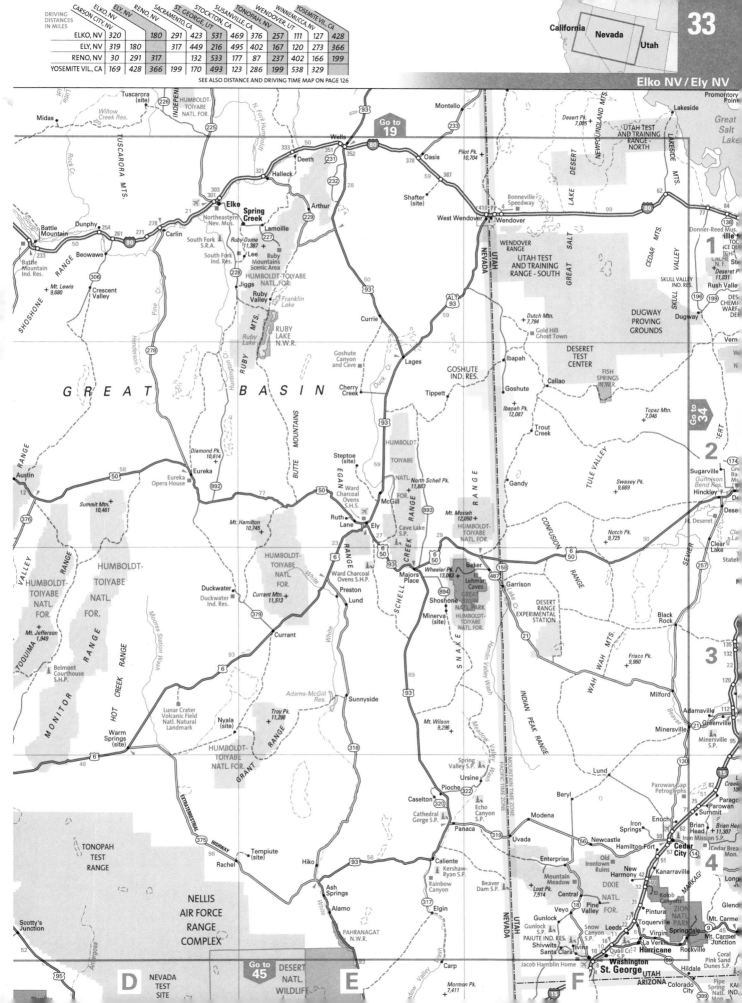

DRIVING DISTANCES IN MILES	CARSON CITY, NV	ELKO, NV	ELY, NV	RENO, NV	SACRAMENTO, CA	ST. GEORGE, UT	STOCKTON, CA	SUSANVILLE, CA	TONOPAH, NV	WENDOVER, NV	WINNEMUCCA, NV	YOSEMITE VIL, CA
ELKO, NV	320		180	291	423	531	469	376	257	111	127	428
ELY, NV	319	180		317	449	216	495	402	167	120	273	366
RENO, NV	30	291	317		132	533	177	87	237	402	166	199
YOSEMITE VIL, CA	169	428	366	199	170	493	123	286	199	538	329	

SEE ALSO DISTANCE AND DRIVING TIME MAP ON PAGE 126

Go to 19

Go to 34

Go to 45

Go to DESERT
NATL.
WILDLIFE

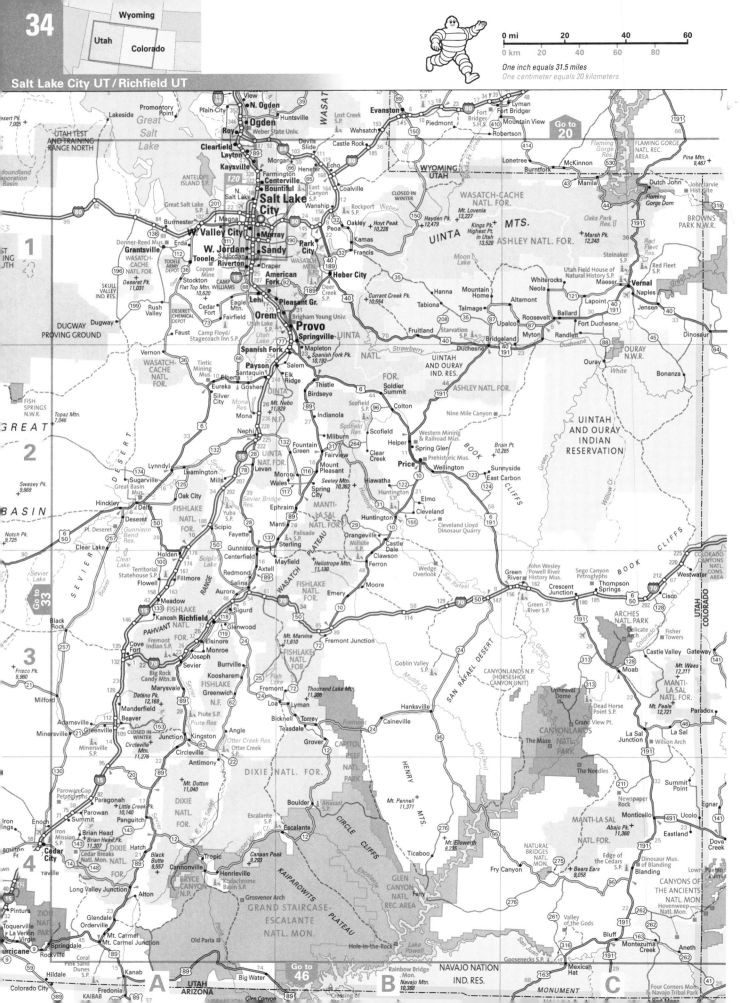

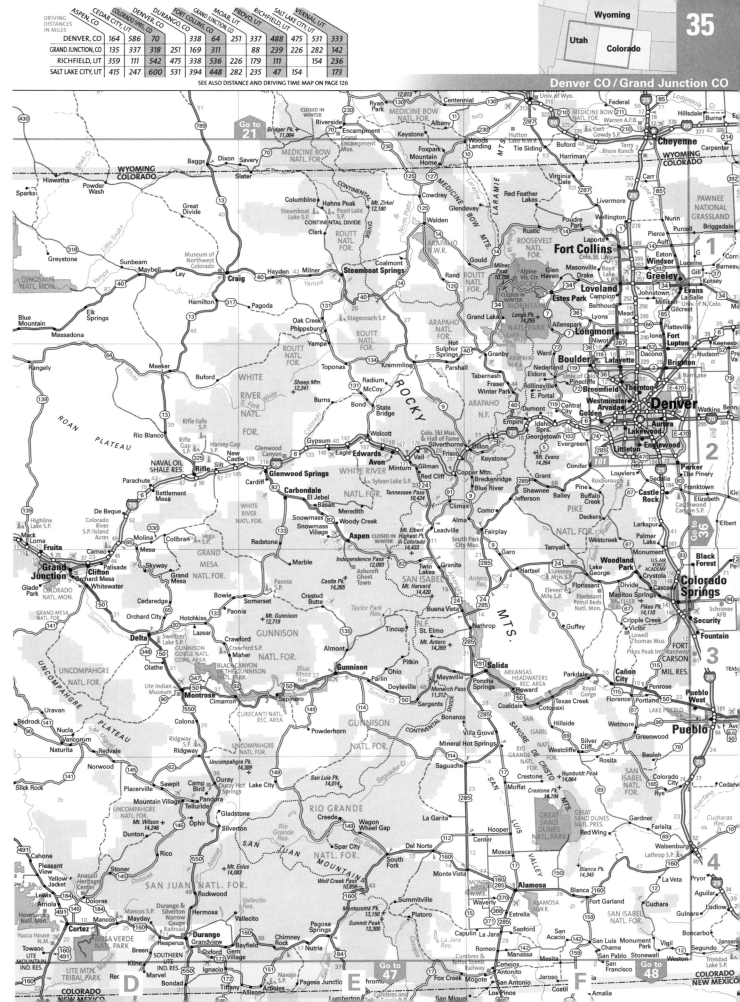

DRIVING DISTANCES IN MILES	ASPEN, CO	CEDAR CITY, UT	COLORADO SPRS, CO	DENVER, CO	DURANGO, CO	FORT COLLINS, CO	GRAND JUNCTION, CO	MOAB, UT	PROVO, UT	RICHFIELD, UT	SALT LAKE CITY, UT	VERNAL, UT
DENVER, CO	164	586	70		338	64	251	337	488	475	531	333
GRAND JUNCTION, CO	135	337	318	251	169	311		88	239	226	282	142
RICHFIELD, UT	359	111	542	475	338	536	226	179	111		154	236
SALT LAKE CITY, UT	415	247	600	531	394	448	282	235	47	154		173

SEE ALSO DISTANCE AND DRIVING TIME MAP ON PAGE 126

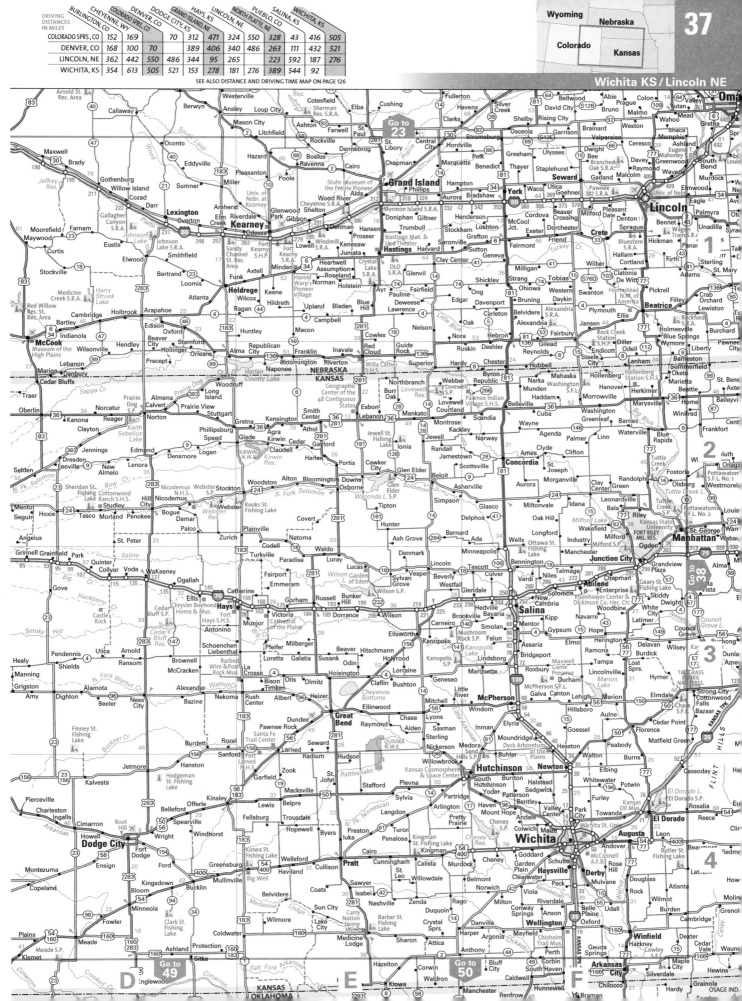

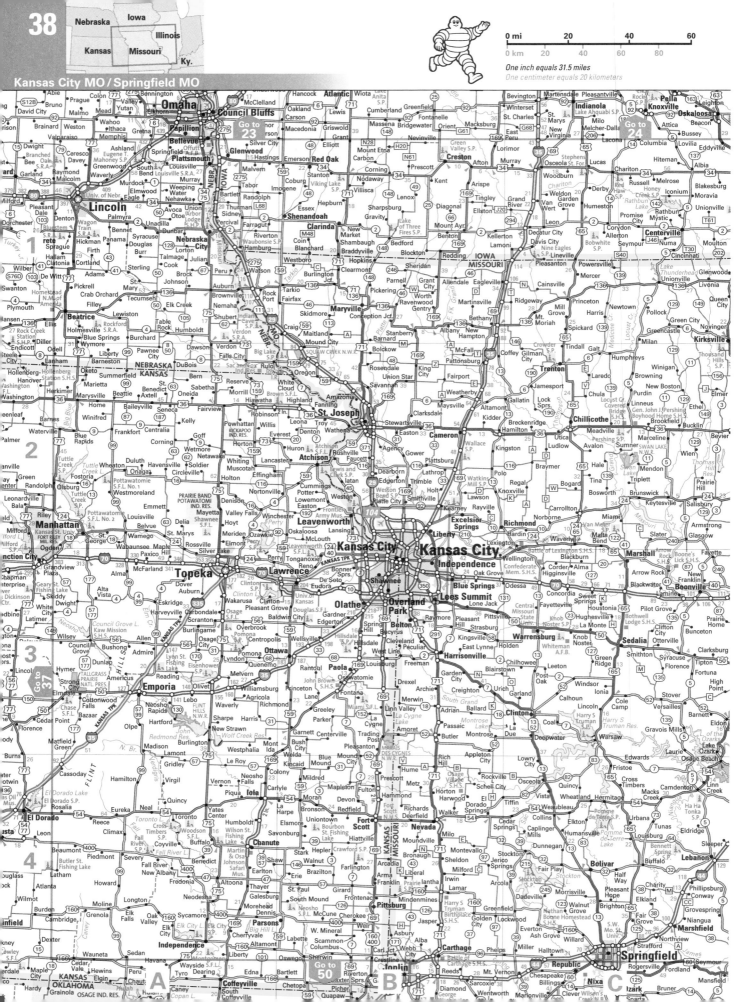

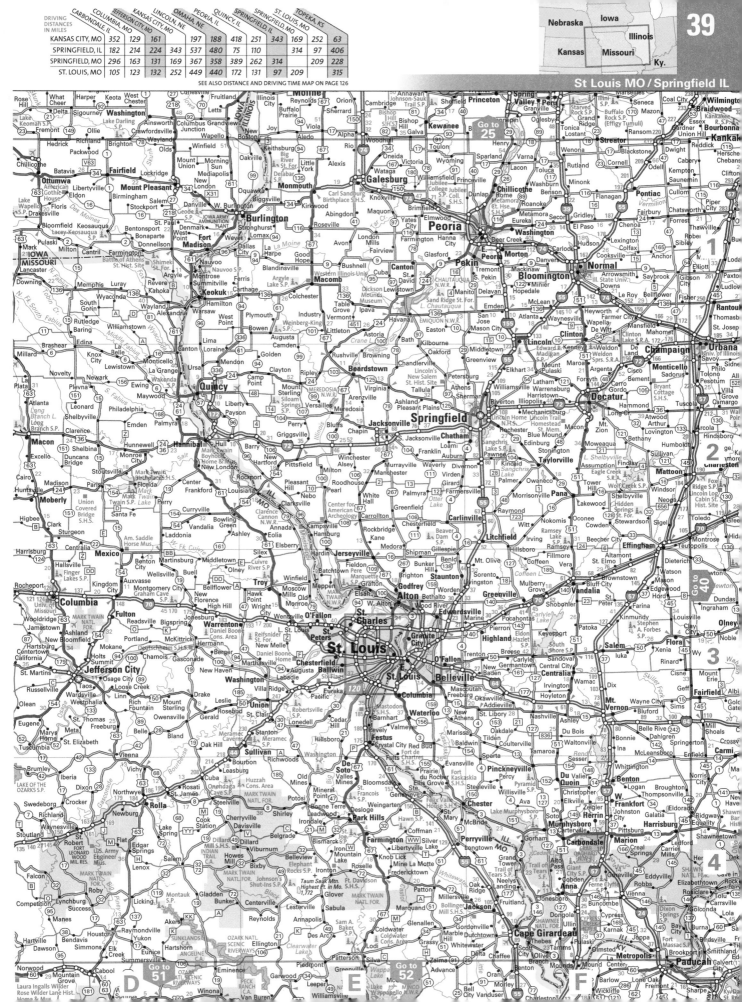

DRIVING DISTANCES IN MILES	CARBONDALE, IL	COLUMBIA, MO	JEFFERSON CITY, MO	KANSAS CITY, MO	LINCOLN, NE	OMAHA, NE	PEORIA, IL	QUINCY, IL	SPRINGFIELD, IL	SPRINGFIELD, MO	ST. LOUIS, MO	TOPEKA, KS
KANSAS CITY, MO	352	129	161		197	188	418	251	343	169	252	63
SPRINGFIELD, IL	182	214	224	343	537	480	75	110		314	97	406
SPRINGFIELD, MO	296	163	131	169	367	358	389	262	314		209	228
ST. LOUIS, MO	105	123	132	252	449	440	172	131	97	209		315

SEE ALSO DISTANCE AND DRIVING TIME MAP ON PAGE 126

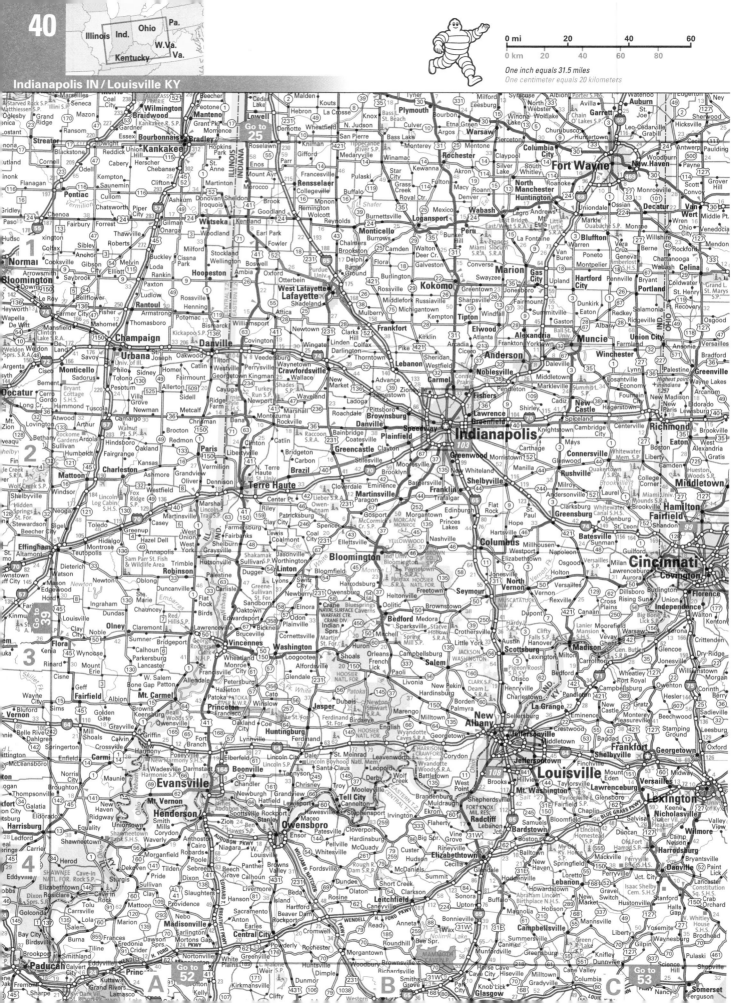

DRIVING DISTANCES IN MILES

	AKRON, OH	BECKLEY, WV	CHAMPAIGN, IL	CHARLESTON, WV	CINCINNATI, OH	COLUMBUS, OH	EVANSVILLE, IN	FORT WAYNE, IN	INDIANAPOLIS, IN	LEXINGTON, KY	LOUISVILLE, KY	WHEELING, WV
AKRON, OH		274	435	214	243	129	452	237	304	326	341	116
COLUMBUS, OH	129	228	307	168	109		319	186	176	193	207	130
INDIANAPOLIS, IN	304	380	123	320	116	176	166	128		191	112	310
LOUISVILLE, KY	341	311	237	251	100	207	114	236	112	80		333

SEE ALSO DISTANCE AND DRIVING TIME MAP ON PAGE 126

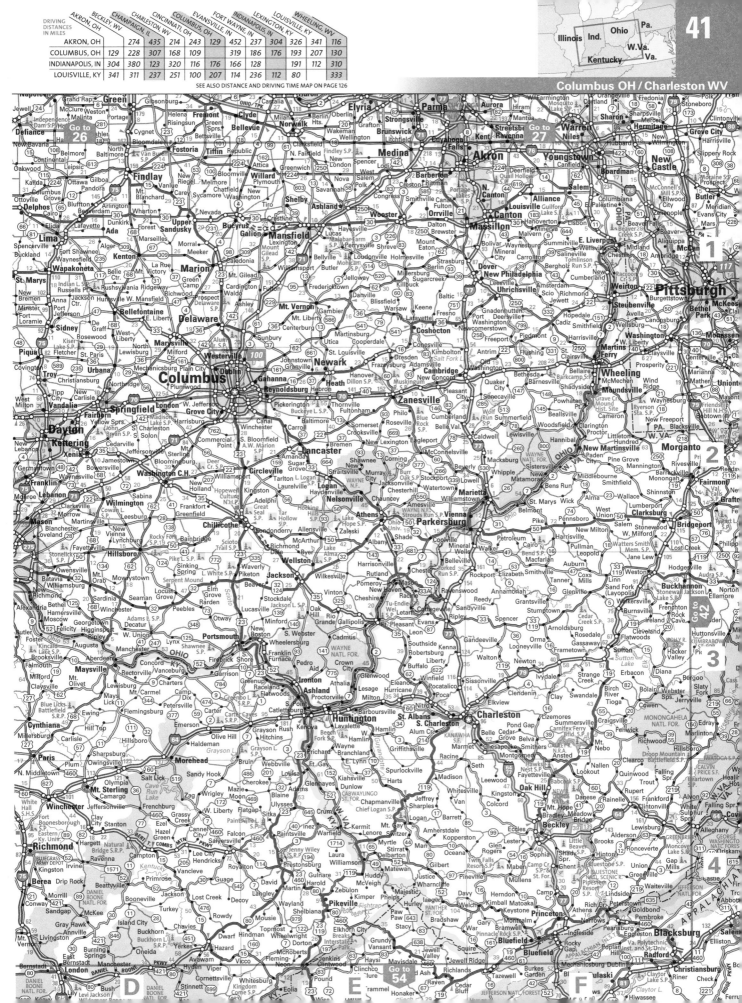

0 mi   20   40   60
0 km   20   40   60   80

One inch equals 31.5 miles
One centimeter equals 20 kilometers

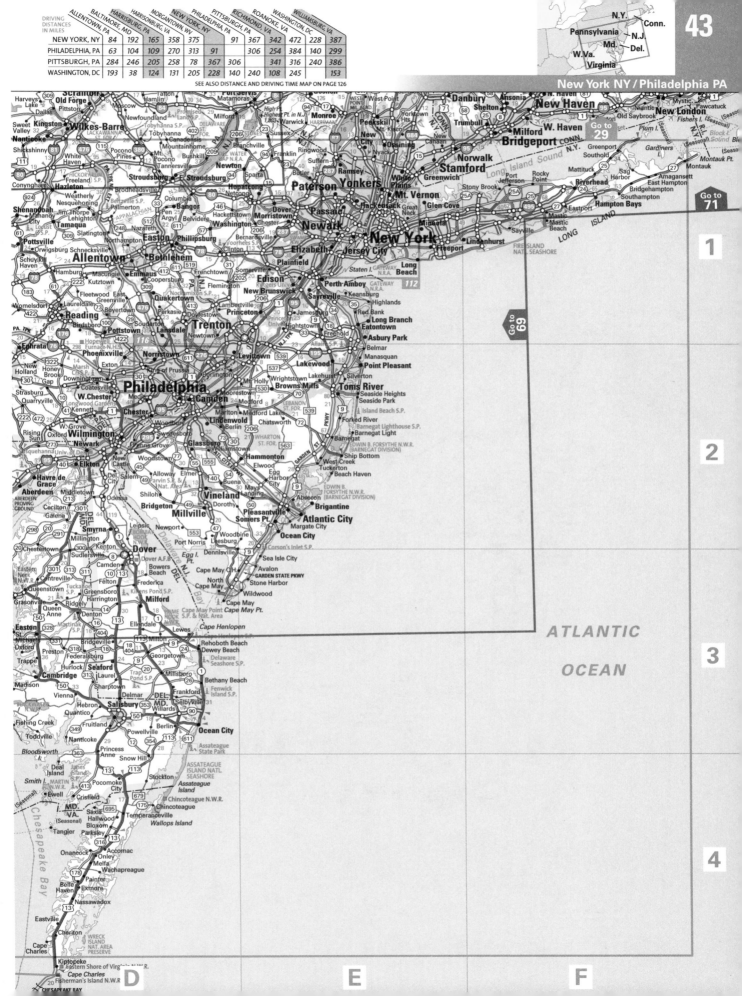

DRIVING DISTANCES IN MILES	ALLENTOWN/N. PA	BALTIMORE, MD	HARRISBURG, PA	HARRISONBURG, VA	MORGANTOWN, WV	NEW YORK, NY	PHILADELPHIA, PA	PITTSBURGH, PA	RICHMOND, VA	ROANOKE, VA	WASHINGTON, DC	WILLIAMSBURG, VA
NEW YORK, NY	84	192	165	358	375		91	367	342	472	228	387
PHILADELPHIA, PA	63	104	109	270	313	91		306	254	384	140	299
PITTSBURGH, PA	284	246	205	258	78	367	306		341	316	240	386
WASHINGTON, DC	193	38	124	131	205	228	140	240	108	245		153

SEE ALSO DISTANCE AND DRIVING TIME MAP ON PAGE 126

DRIVING DISTANCES IN MILES

	BAKERSFIELD, CA	BARSTOW, CA	BLYTHE, CA	FRESNO, CA	KINGMAN, AZ	LAS VEGAS, NV	LOS ANGELES, CA	PALM SPRINGS, CA	ST. GEORGE, UT	SALINAS, CA	SAN LUIS OBISPO, CA	SANTA BARBARA, CA
BARSTOW, CA	136		246	248	210	156	118	126	274	345	306	213
LAS VEGAS, NV	287	156	211	399	103		274	282	118	497	406	369
LOS ANGELES, CA	111	118	230	219	328	274		110	392	307	190	97
SALINAS, CA	209	345	532	145	555	497	307	412	615		130	222

SEE ALSO DISTANCE AND DRIVING TIME MAP ON PAGE 126

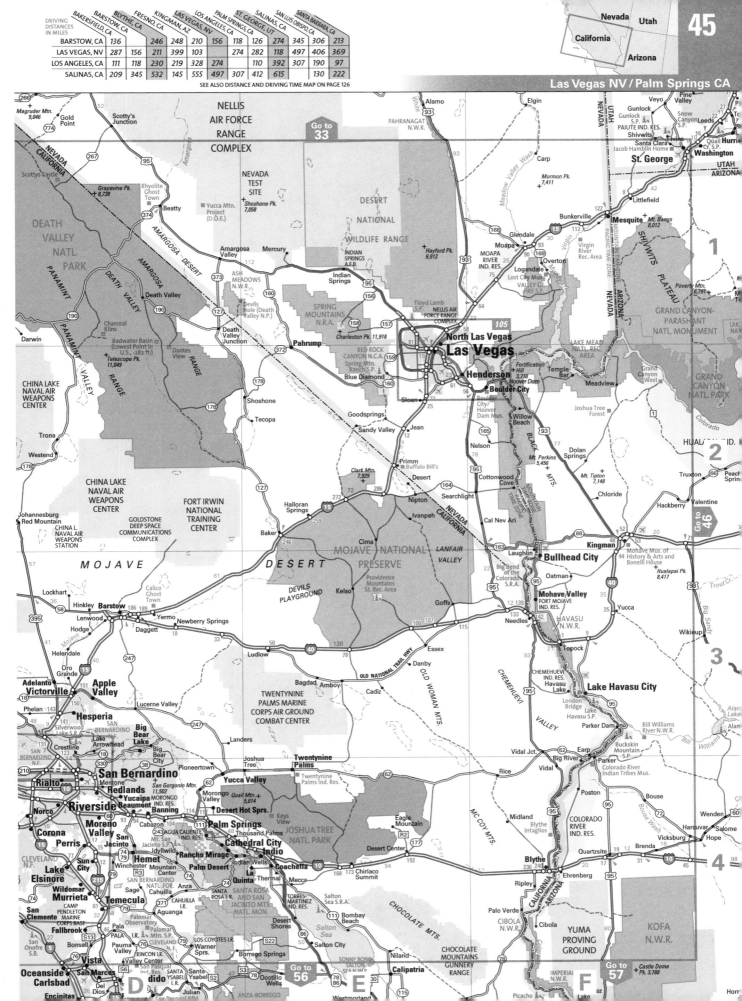

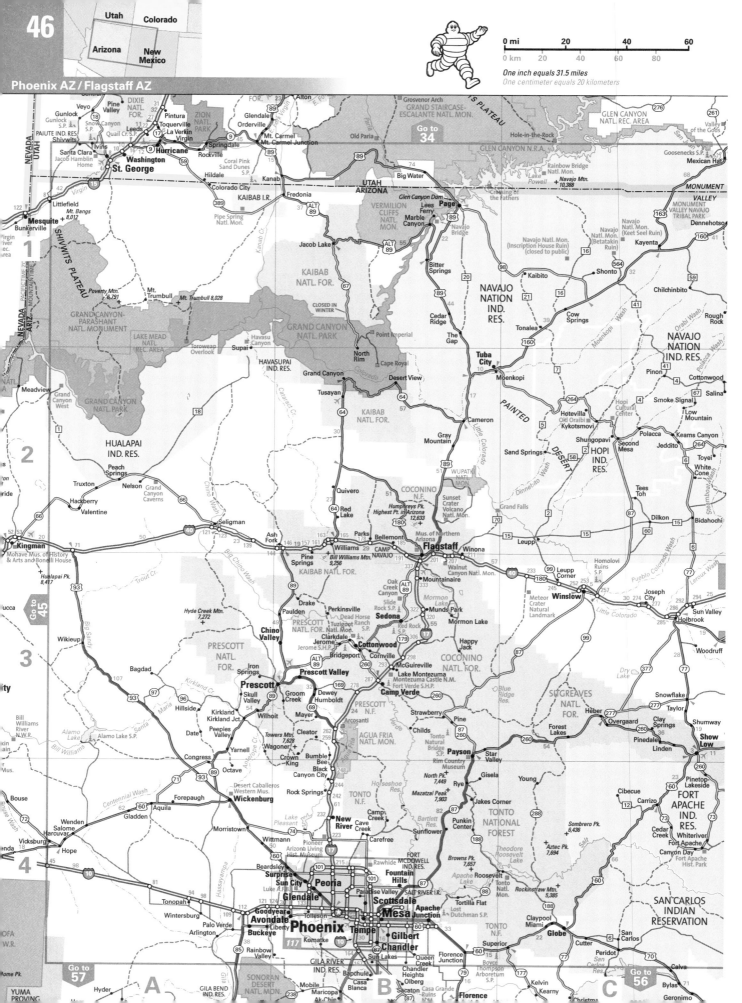

Utah | Colorado
Arizona | New Mexico

0 mi   20   40   60
0 km   20   40   60   80
One inch equals 31.5 miles
One centimeter equals 20 kilometers

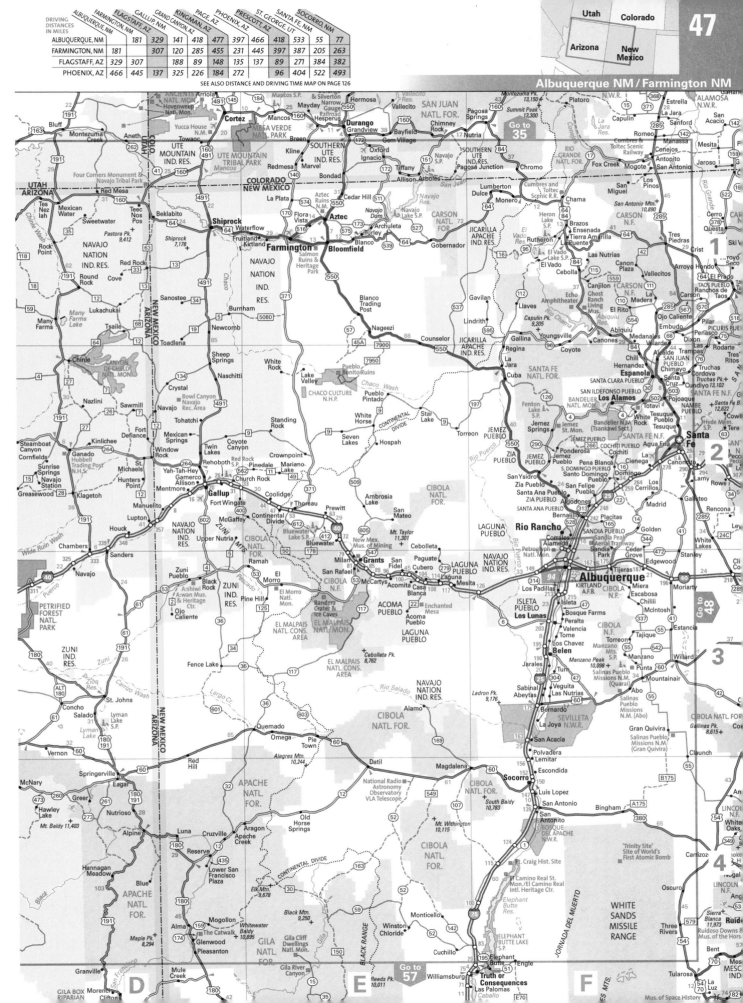

DRIVING DISTANCES IN MILES	ALBUQUERQUE, NM	FARMINGTON, NM	FLAGSTAFF, AZ	GALLUP, NM	GRAND CANYON, AZ	KINGMAN, AZ	PAGE, AZ	PHOENIX, AZ	PRESCOTT, AZ	ST. GEORGE, UT	SANTA FE, NM	SOCORRO, NM
ALBUQUERQUE, NM		181	329	141	418	477	397	466	418	533	55	77
FARMINGTON, NM	181		307	120	285	455	231	445	397	387	205	263
FLAGSTAFF, AZ	329	307		188	89	148	135	137	89	271	384	382
PHOENIX, AZ	466	445	137	325	184	272		96	404	522	493

SEE ALSO DISTANCE AND DRIVING TIME MAP ON PAGE 126

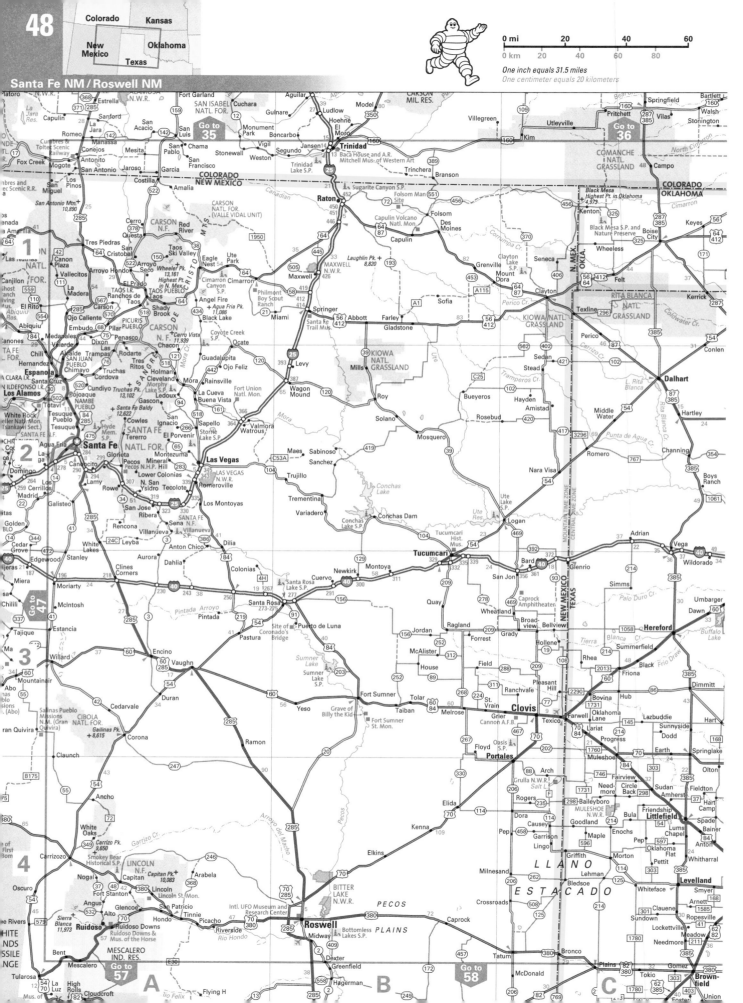

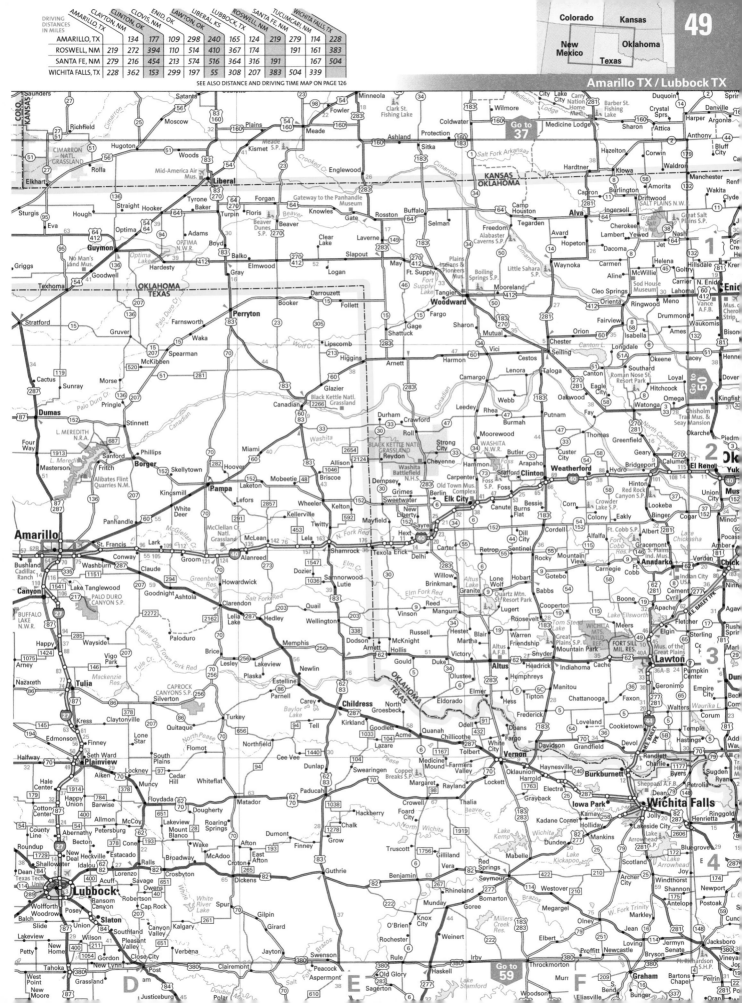

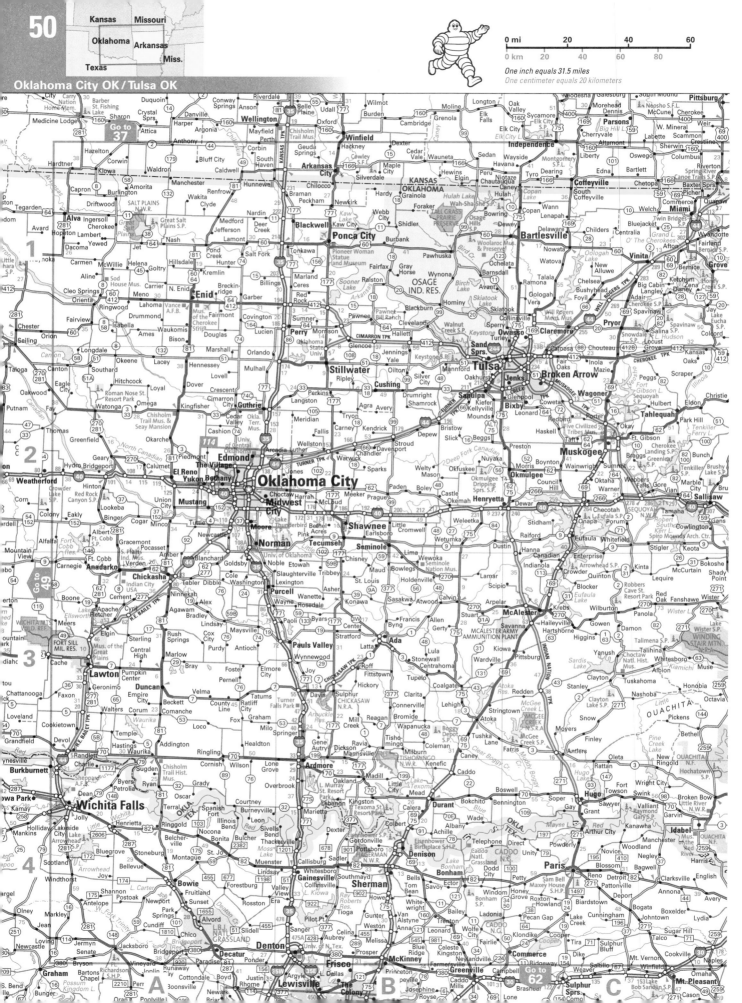

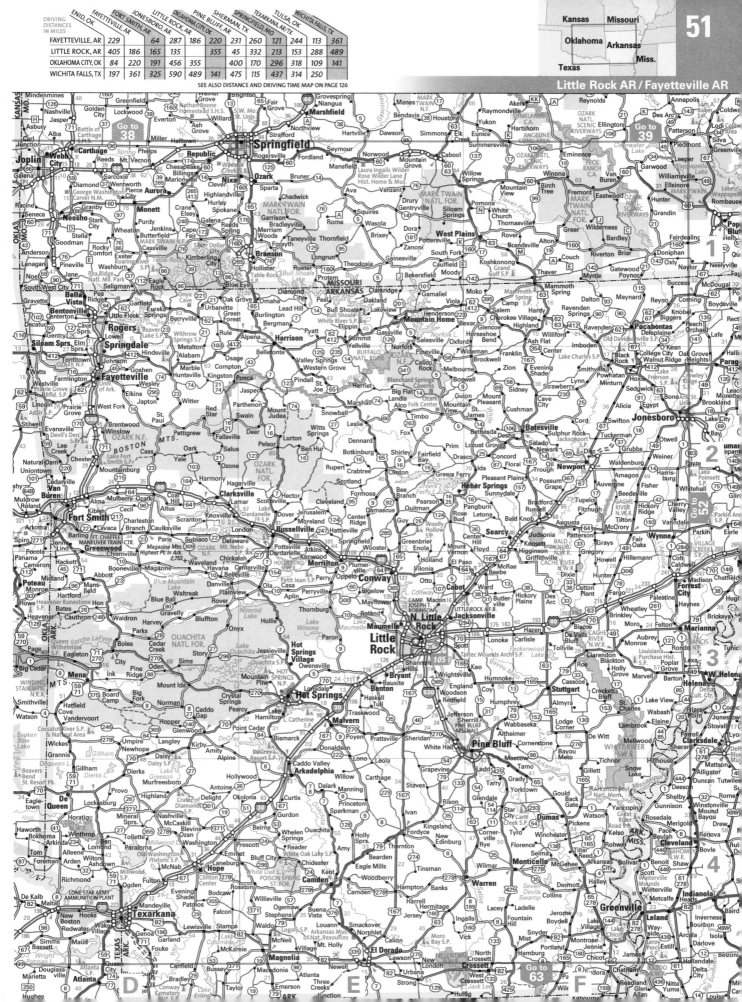

SEE ALSO DISTANCE AND DRIVING TIME MAP ON PAGE 126

DRIVING DISTANCES IN MILES	ENID, OK	FAYETTEVILLE, AR	FORT SMITH, AR	JONESBORO, AR	LITTLE ROCK, AR	OKLAHOMA CITY, OK	PINE BLUFF, AR	SHERMAN, TX	SPRINGFIELD, MO	TEXARKANA, AR/TX	TULSA, OK	WICHITA FALLS, TX
FAYETTEVILLE, AR	229		64	287	186	220	231	260	121	244	113	361
LITTLE ROCK, AR	405	186	165	135		355	45	332	213	153	288	489
OKLAHOMA CITY, OK	84	220	191	456	355		400	170	296	318	109	141
WICHITA FALLS, TX	197	361	325	590	489	141		475	437	314	250	

Kansas   Missouri

Oklahoma   Arkansas

Texas   Miss.

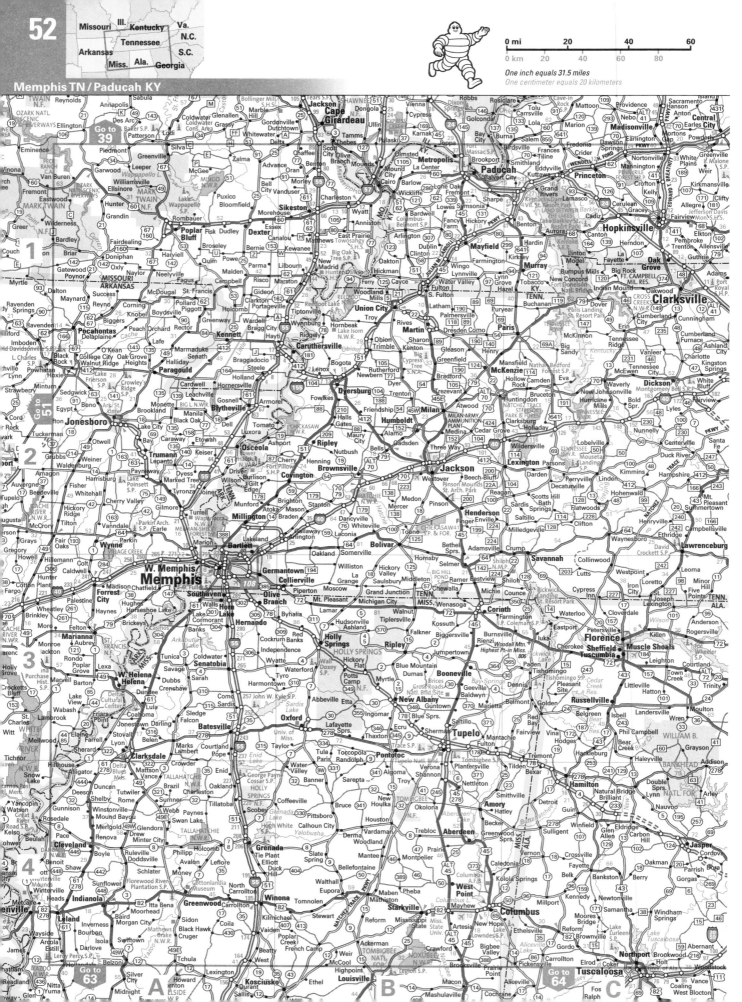

0 mi   20   40   60

0 km   20   40   60   80

One inch equals 31.5 miles
One centimeter equals 20 kilometers

Go to 39
Go to 51
Go to 63
Go to 64

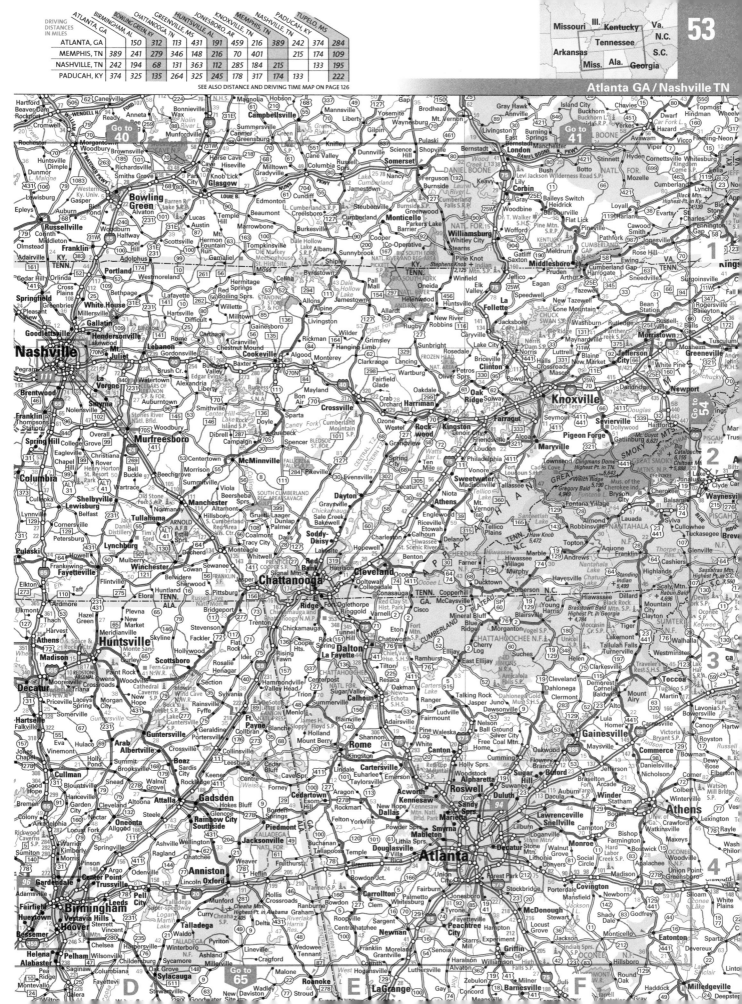

DRIVING DISTANCES IN MILES	ATLANTA, GA	BIRMINGHAM, AL	BOWLING GREEN, KY	CHATTANOOGA, TN	GREENVILLE, MS	HUNTSVILLE, AL	JONESBORO, AR	KNOXVILLE, TN	MEMPHIS, TN	NASHVILLE, TN	PADUCAH, KY	TUPELO, MS		
ATLANTA, GA		150	312	113	431	191	459	216	389	242	374	284		
MEMPHIS, TN	389	241	279	346	148	216	70	401		215	174	109		
NASHVILLE, TN	242	194	68	131	363	112	285	184	215		133	195		
PADUCAH, KY	374	325	135	264	325	245	178	317	174	133		222		

SEE ALSO DISTANCE AND DRIVING TIME MAP ON PAGE 126

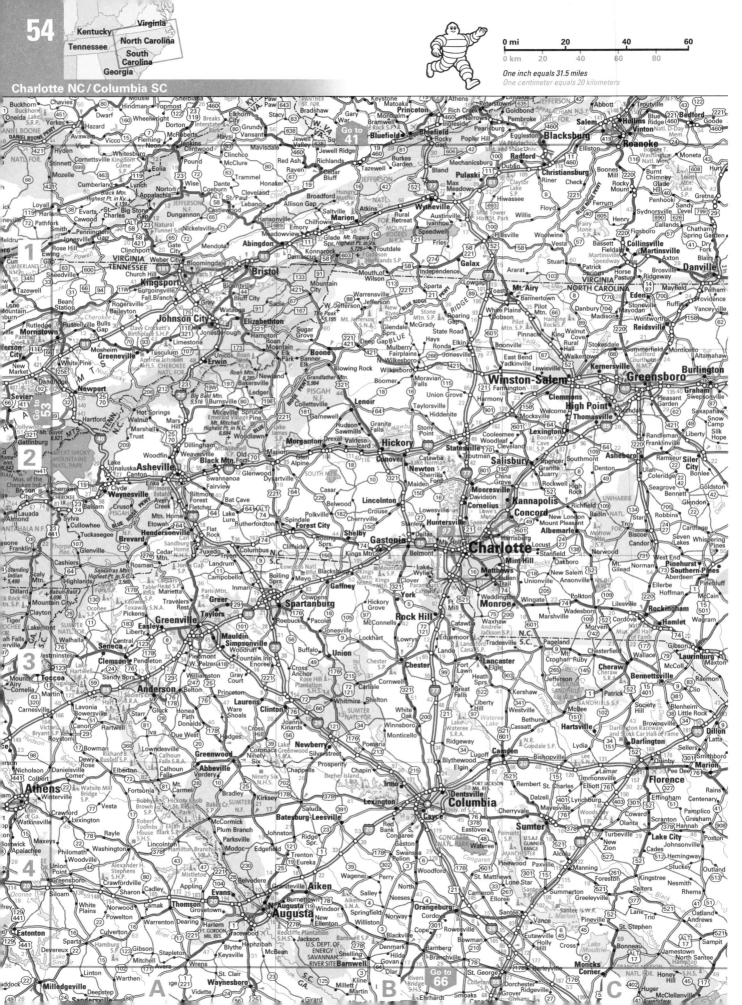

Kentucky   Virginia
         North Carolina
Tennessee   South
              Carolina
     Georgia

0 mi   20   40   60
0 km   20   40   60   80
One inch equals 31.5 miles
One centimeter equals 20 kilometers

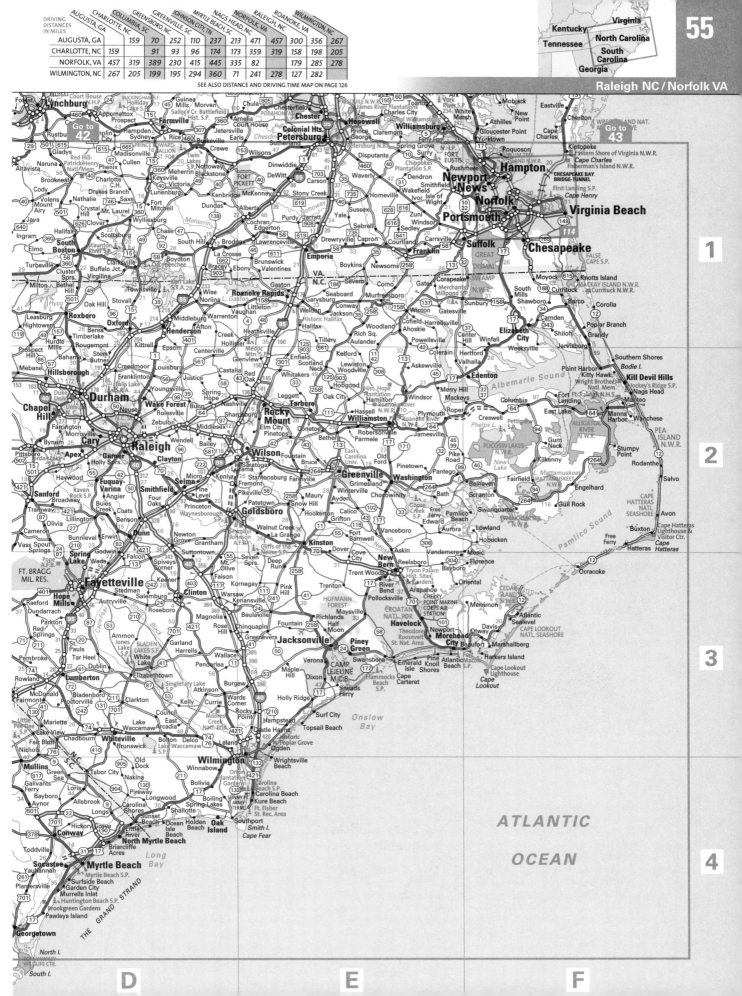

DRIVING DISTANCES IN MILES	AUGUSTA, GA	CHARLOTTE, NC	COLUMBIA, SC	GREENSBORO, NC	GREENVILLE, SC	JOHNSON CITY, TN	MYRTLE BEACH, SC	NAGS HEAD, NC	NORFOLK, VA	RALEIGH, NC	ROANOKE, VA	WILMINGTON, NC
AUGUSTA, GA		159	70	252	110	237	213	471	457	300	356	267
CHARLOTTE, NC	159		91	93	96	174	173	359	319	158	198	205
NORFOLK, VA	457	319	389	230	415	445	335	82		179	285	278
WILMINGTON, NC	267	205	199	195	294	360	71	241	278	127	282	

SEE ALSO DISTANCE AND DRIVING TIME MAP ON PAGE 126

Go to 42

Go to 43

Kentucky
Tennessee
Virginia
North Carolina
South Carolina
Georgia

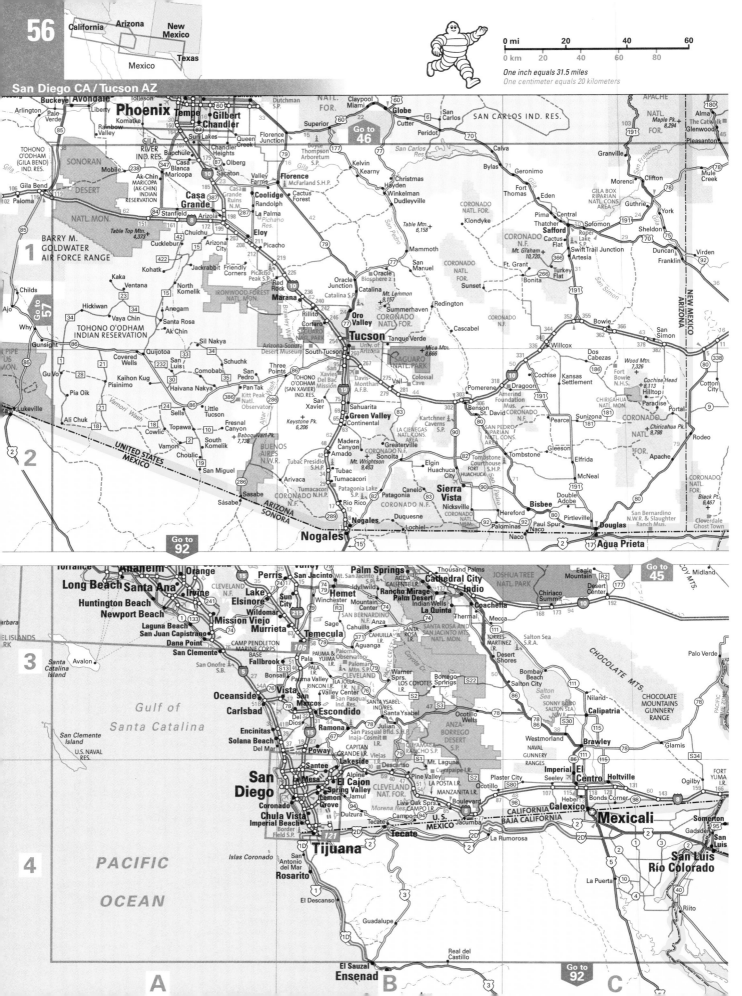

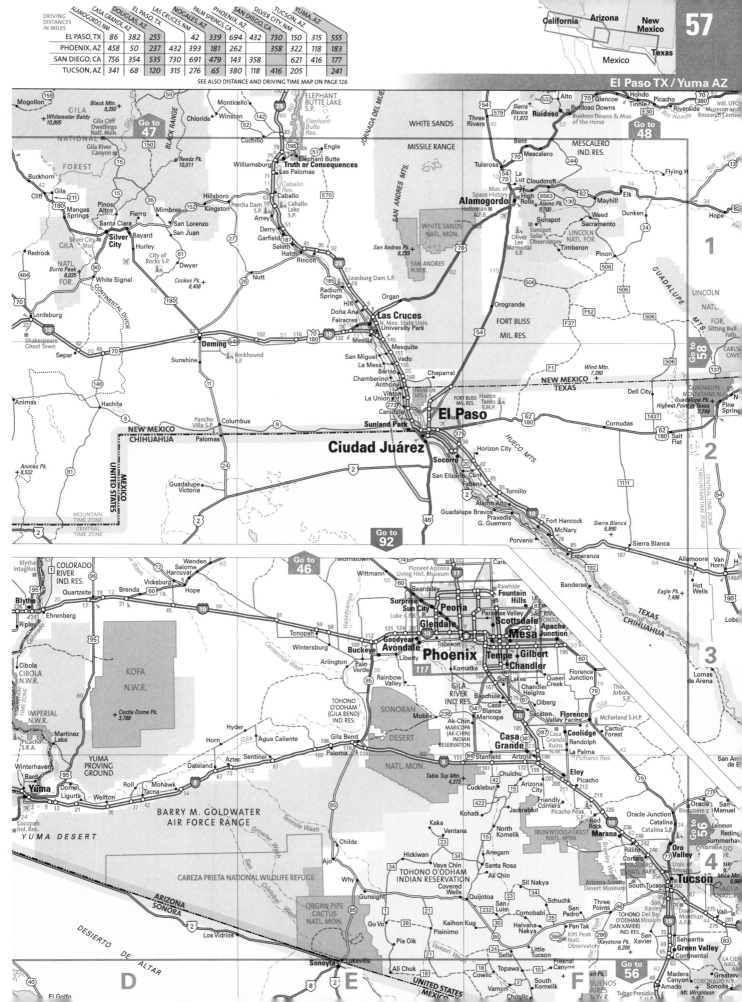

DRIVING DISTANCES IN MILES

	ALAMOGORDO, NM	CASA GRANDE, AZ	DOUGLAS, AZ	EL PASO, TX	LAS CRUCES, NM	NOGALES, AZ	PALM SPRINGS, CA	PHOENIX, AZ	SAN DIEGO, CA	SILVER CITY, NM	TUCSON, AZ	YUMA, AZ
EL PASO, TX	86	382	255		42	339	694	432	730	150	315	555
PHOENIX, AZ	458	50	237	432	393	181	262		358	322	118	183
SAN DIEGO, CA	756	354	535	730	691	479	143	358		621	416	177
TUCSON, AZ	341	68	120	315	276	65	380	118	416	205		241

SEE ALSO DISTANCE AND DRIVING TIME MAP ON PAGE 126

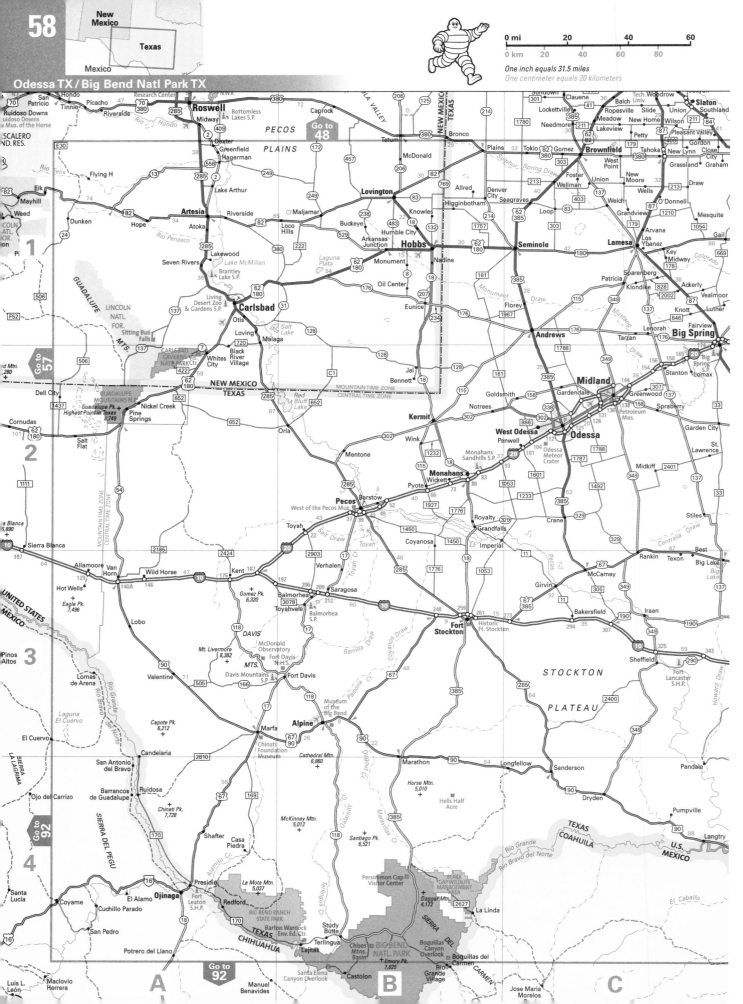

0 mi   20   40   60
0 km 20 40 60 80
One inch equals 31.5 miles
One centimeter equals 20 kilometers

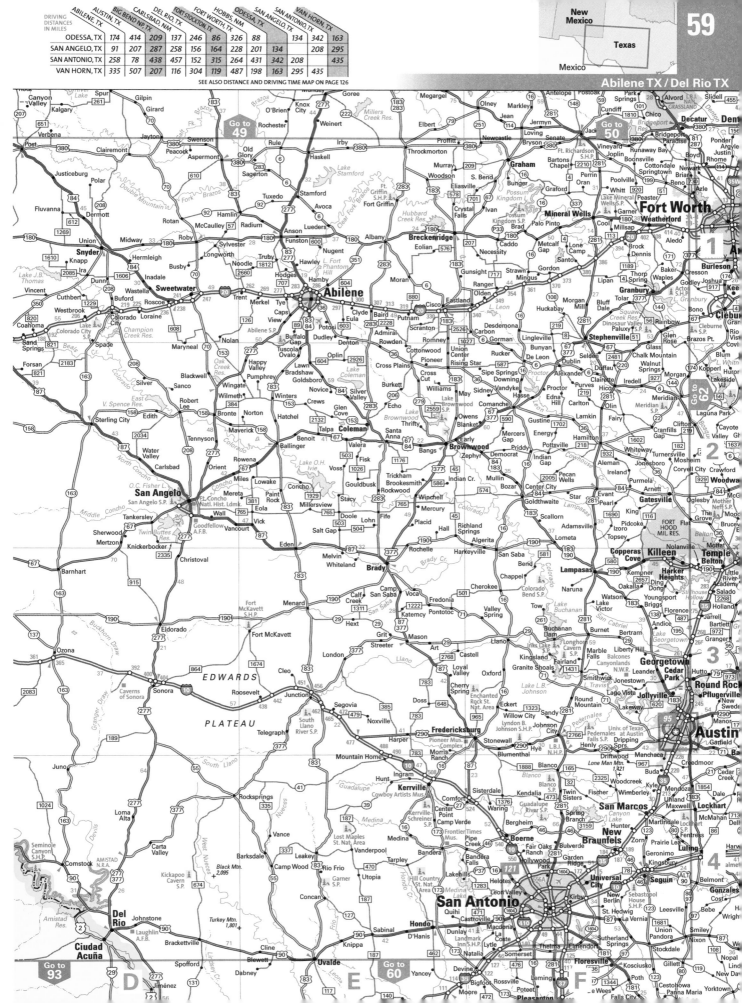

DRIVING
DISTANCES
IN MILES

	ABILENE, TX	AUSTIN, TX	BIG BEND NP TX	CARLSBAD, NM	DEL RIO, TX	FORT STOCKTON, TX	FORT WORTH, TX	HOBBS, NM	ODESSA, TX	SAN ANGELO, TX	SAN ANTONIO, TX	VAN HORN, TX
ODESSA, TX	174	414	209	137	246	86	326	88		134	342	163
SAN ANGELO, TX	91	207	287	258	156	164	228	201	134		208	295
SAN ANTONIO, TX	258	78	438	457	152	264	431	342	342	208		435
VAN HORN, TX	335	507	207	116	304	119	487	198	163	295	435	

SEE ALSO DISTANCE AND DRIVING TIME MAP ON PAGE 126

Go to 49

Go to 50

Go to 62

Go to 93

Go to 60

New Mexico

Texas

Mexico

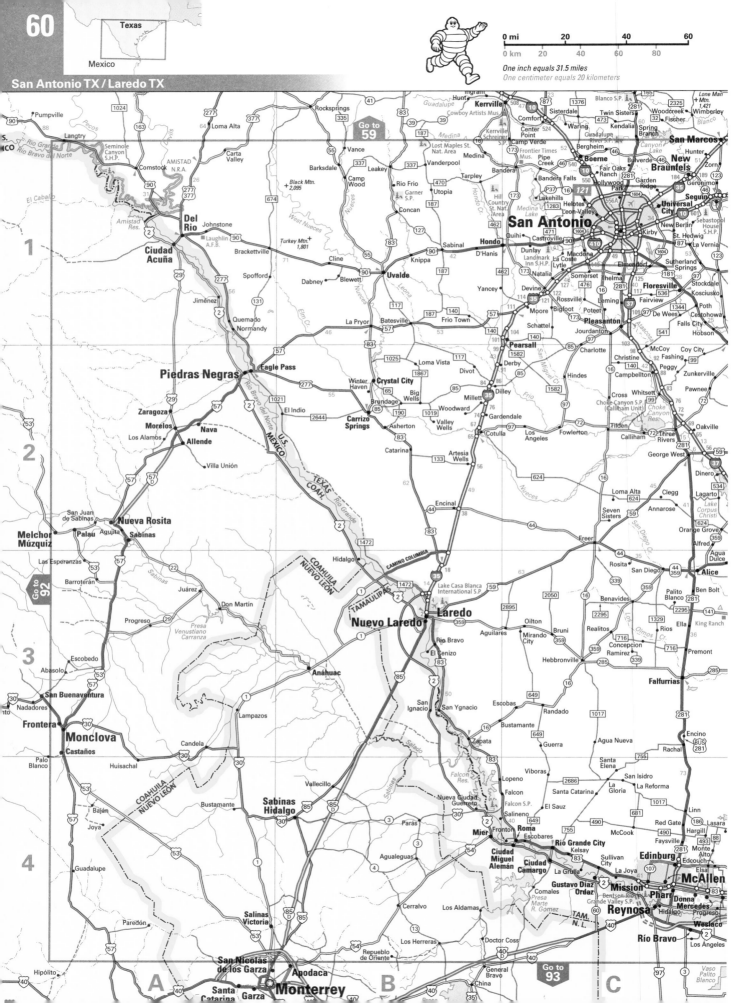

0 mi   20   40   60

0 km   20   40   60   80

One inch equals 31.5 miles
One centimeter equals 20 kilometers

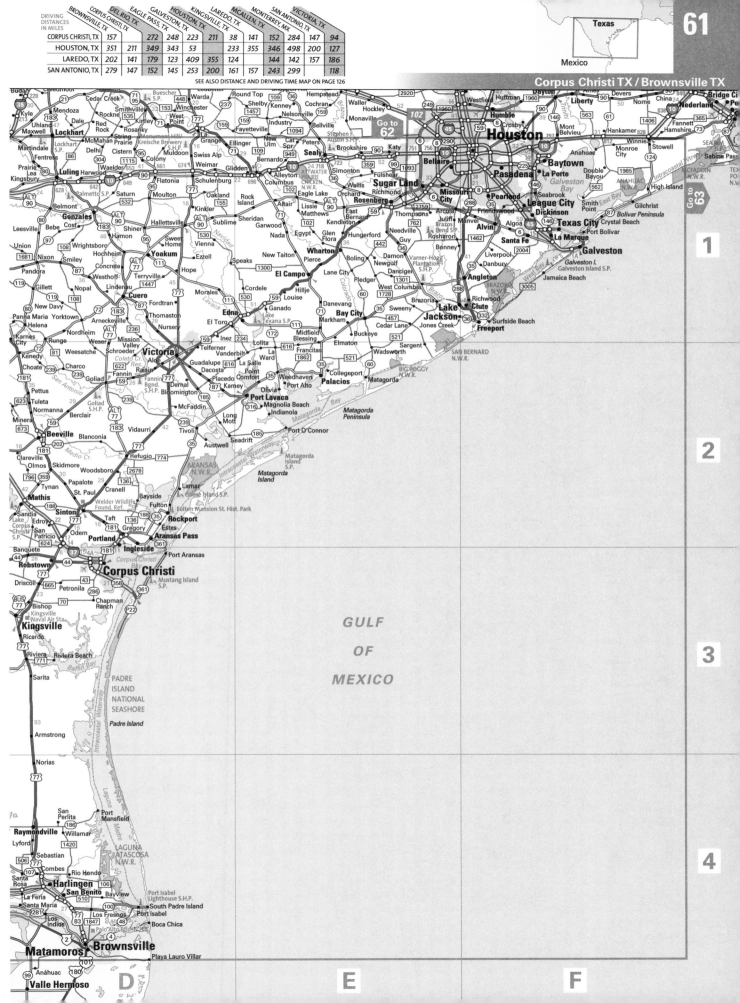

DRIVING DISTANCES IN MILES	BROWNSVILLE, TX	CORPUS CHRISTI, TX	DEL RIO, TX	EAGLE PASS, TX	GALVESTON, TX	HOUSTON, TX	KINGSVILLE, TX	LAREDO, TX	McALLEN, TX	MONTERREY, MX	SAN ANTONIO, TX	VICTORIA, TX
CORPUS CHRISTI, TX	157		272	248	223	211	38	141	152	284	147	94
HOUSTON, TX	351	211	349	343	53		233	355	346	498	200	127
LAREDO, TX	202	141	179	123	409	355	124		144	142	157	186
SAN ANTONIO, TX	279	147	152	145	253	200	161	157	243	299		118

SEE ALSO DISTANCE AND DRIVING TIME MAP ON PAGE 126

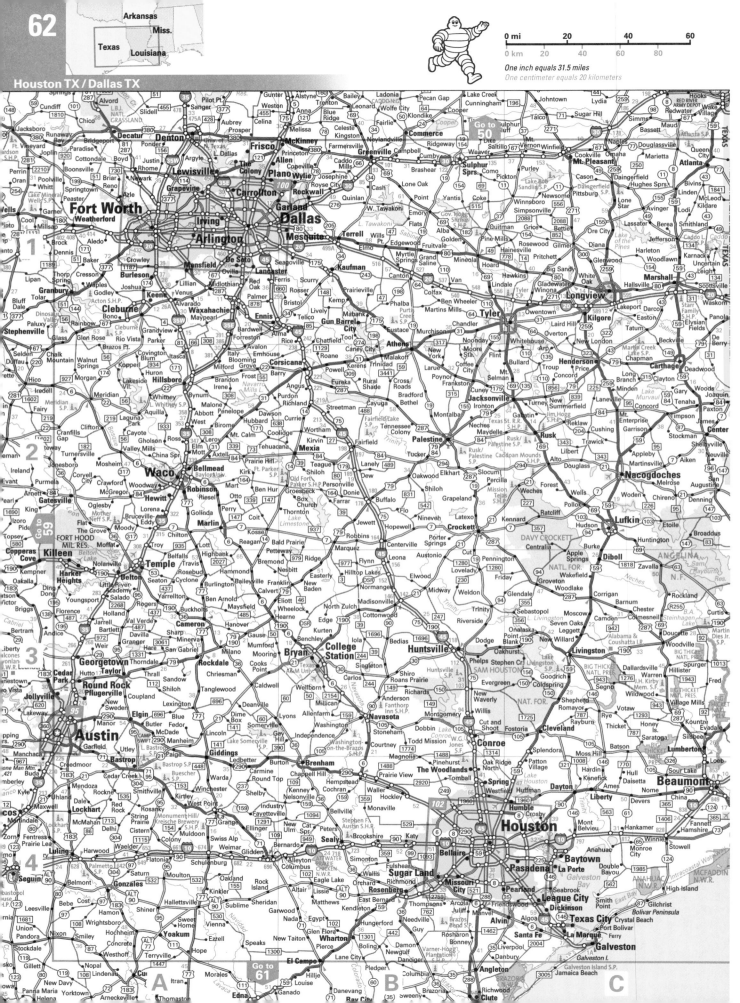

Arkansas
Miss.
Texas
Louisiana

0 mi   20   40   60
0 km   20   40   60   80
One inch equals 31.5 miles
One centimeter equals 20 kilometers

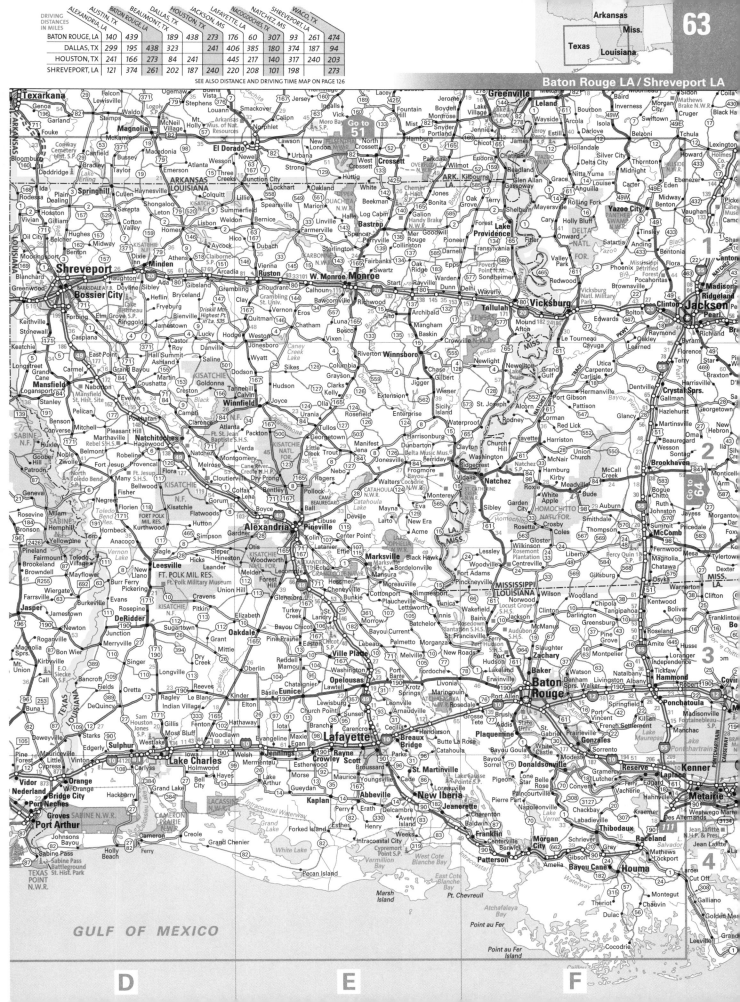

SEE ALSO DISTANCE AND DRIVING TIME MAP ON PAGE 126

DRIVING DISTANCES IN MILES	AUSTIN, TX	BATON ROUGE, LA	BEAUMONT, TX	DALLAS, TX	HOUSTON, TX	JACKSON, MS	LAFAYETTE, LA	NACOGDOCHES, TX	NATCHEZ, MS	SHREVEPORT, LA	WACO, TX
BATON ROUGE, LA	140	439		189	438	273	176	60	307	93	261	474
DALLAS, TX	299	195	438		323		241	406	385	180	374	187	94
HOUSTON, TX	241	166	273	84	241		217	140	317	240	203
SHREVEPORT, LA	121	374	261	202	187	240	220	208	101	198		273

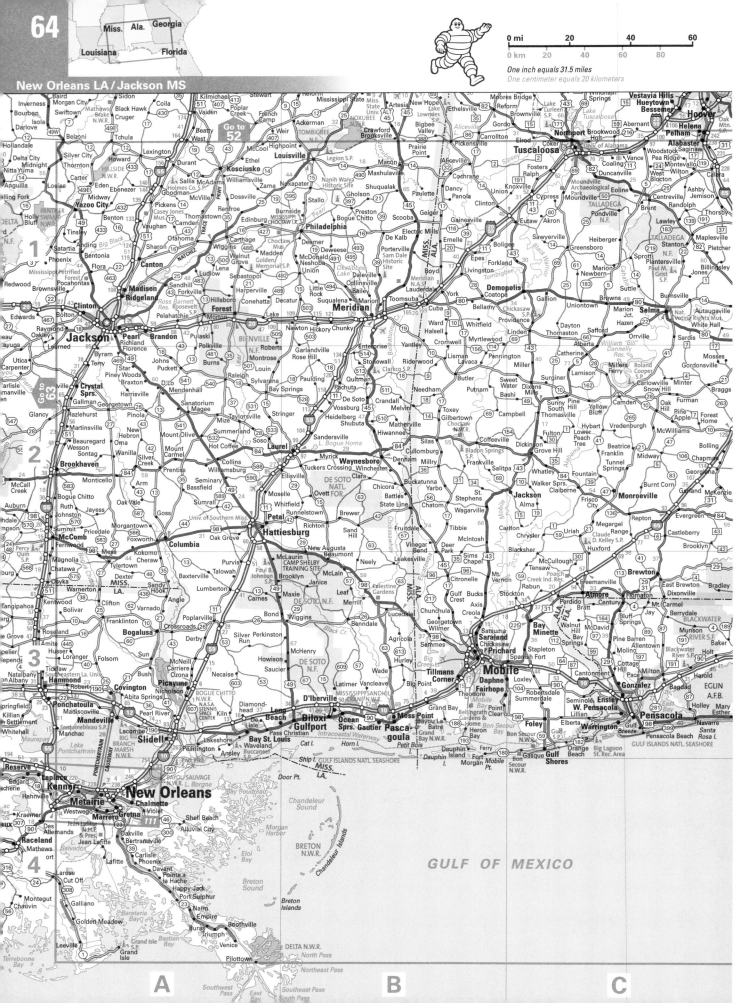

Miss. Ala. Georgia
Louisiana Florida

0 mi 20 40 60
0 km 20 40 60 80
One inch equals 31.5 miles
One centimeter equals 20 kilometers

GULF OF MEXICO

A   B   C

DRIVING DISTANCES IN MILES

	COLUMBUS, GA	DOTHAN, AL	HATTIESBURG, MS	JACKSON, MS	MACON, GA	MERIDIAN, MS	MOBILE, AL	MONTGOMERY, AL	NEW ORLEANS, LA	PANAMA CITY, FL	TALLAHASSEE, FL	TUSCALOOSA, AL
JACKSON, MS	324	338	90		470	91	187	245	185	348	435	185
MONTGOMERY, AL	79	103	243	245	203	153	173		314	180	213	134
NEW ORLEANS, LA	394	345	115	185	517	201	146	314		307	394	295
TALLAHASSEE, FL	169	110	345	435	192	380	247	213	394	104		347

SEE ALSO DISTANCE AND DRIVING TIME MAP ON PAGE 126

DRIVING DISTANCES IN MILES	CHARLESTON, SC	DAYTONA BEACH, FL	FORT MYERS, FL	FORT PIERCE, FL	GAINESVILLE, FL	JACKSONVILLE, FL	KEY WEST, FL	MIAMI, FL	ORLANDO, FL	SARASOTA, FL	SAVANNAH, GA	TAMPA, FL
CHARLESTON, SC		329	533	461	308	238	750	583	379	487	107	434
JACKSONVILLE, FL	238	91	295	223	70		512	345	141	249	141	196
MIAMI, FL	583	260	155	122	338	345	168		229	225	486	274
TAMPA, FL	434	138	123	172	132	196	426	274	82	60	337	

SEE ALSO DISTANCE AND DRIVING TIME MAP ON PAGE 126

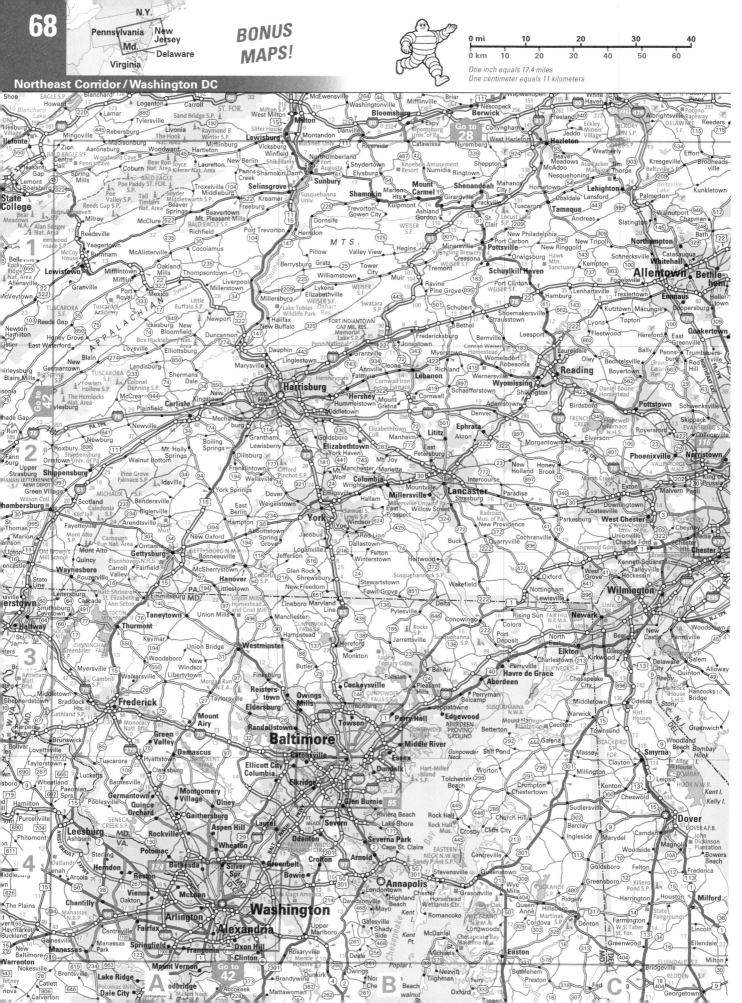

DRIVING DISTANCES IN MILES	ALLENTOWN, PA	ATLANTIC CITY, NJ	BALTIMORE, MD	DOVER, DE	HARRISBURG, PA	LANCASTER, PA	NEWARK, NJ	NEW YORK, NY	PHILADELPHIA, PA	TRENTON, NJ	WASHINGTON, DC	WILMINGTON, DE
HARRISBURG, PA	82	171	83	126		44	154	165	109	135	123	102
NEW YORK, NY	84	125	192	160	165	165	11		91	55	228	120
PHILADELPHIA, PA	63	62	104	74	109	79	80	91		34	140	30
WASHINGTON, DC	188	186	38	94	123	123	218	228	140	179		110

SEE ALSO DISTANCE AND DRIVING TIME MAP ON PAGE 126

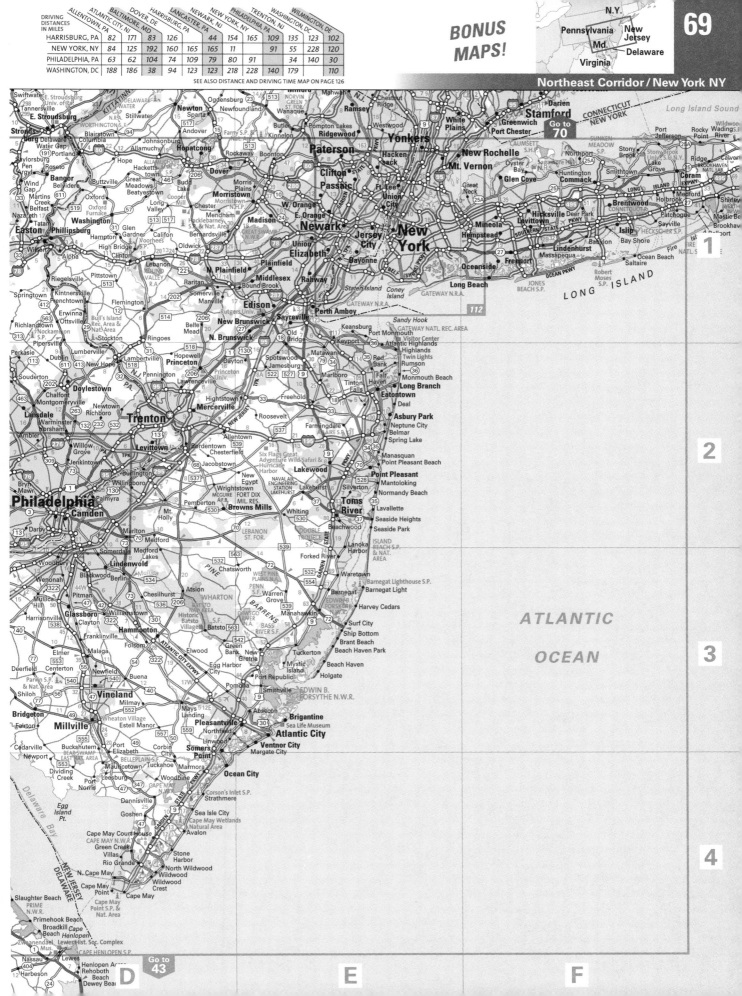

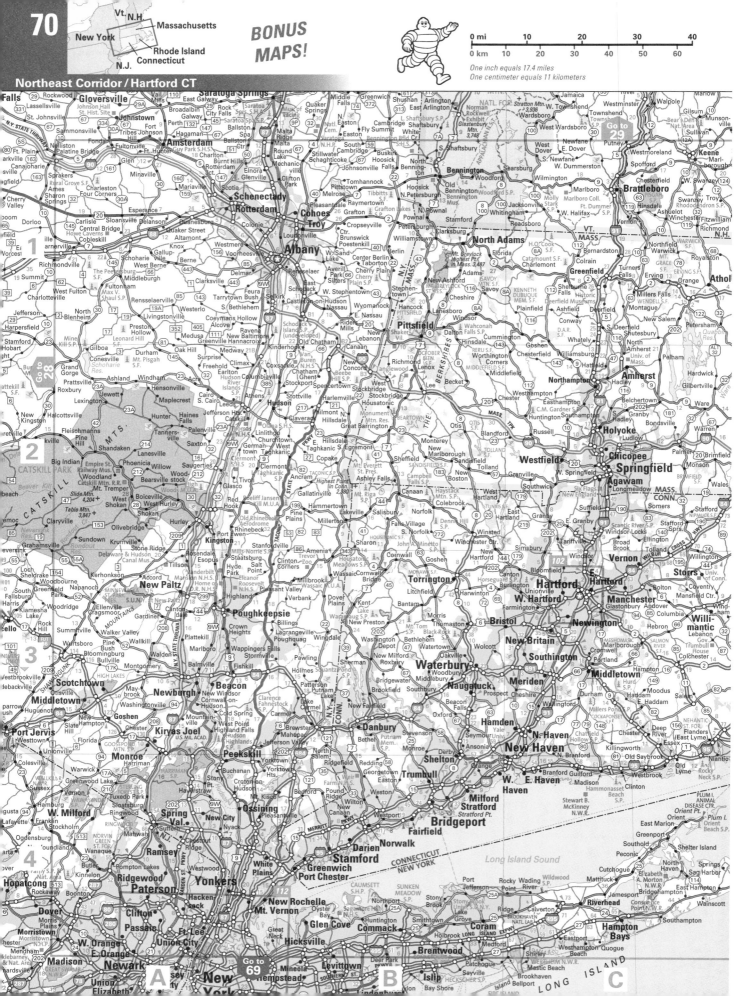

SEE ALSO DISTANCE AND DRIVING TIME MAP ON PAGE 126

DRIVING DISTANCES IN MILES	ALBANY, NY	BOSTON, MA	BRATTLEBORO, VT	HARTFORD, CT	MANCHESTER, NH	NEWBURGH, NY	NEW HAVEN, CT	NEW YORK, NY	PROVIDENCE, RI	PROVINCETOWN, MA	SPRINGFIELD, MA	WORCESTER, MA
ALBANY, NY		172	76	111	145	89	150	151	170	271	86	133
BOSTON, MA	172		108	102	54	201	139	215	52	117	95	46
HARTFORD, CT	111	102	84		131	99	39	115	73	200	25	62
NEW YORK, NY	151	215	195	115	245	56	78		177	292	141	176

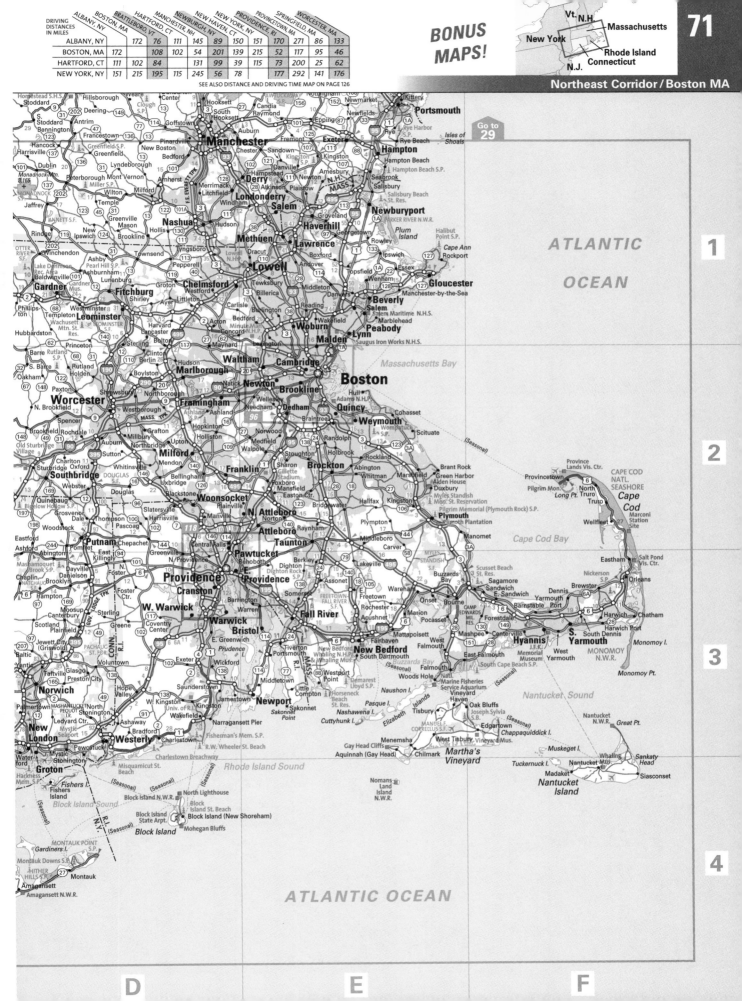

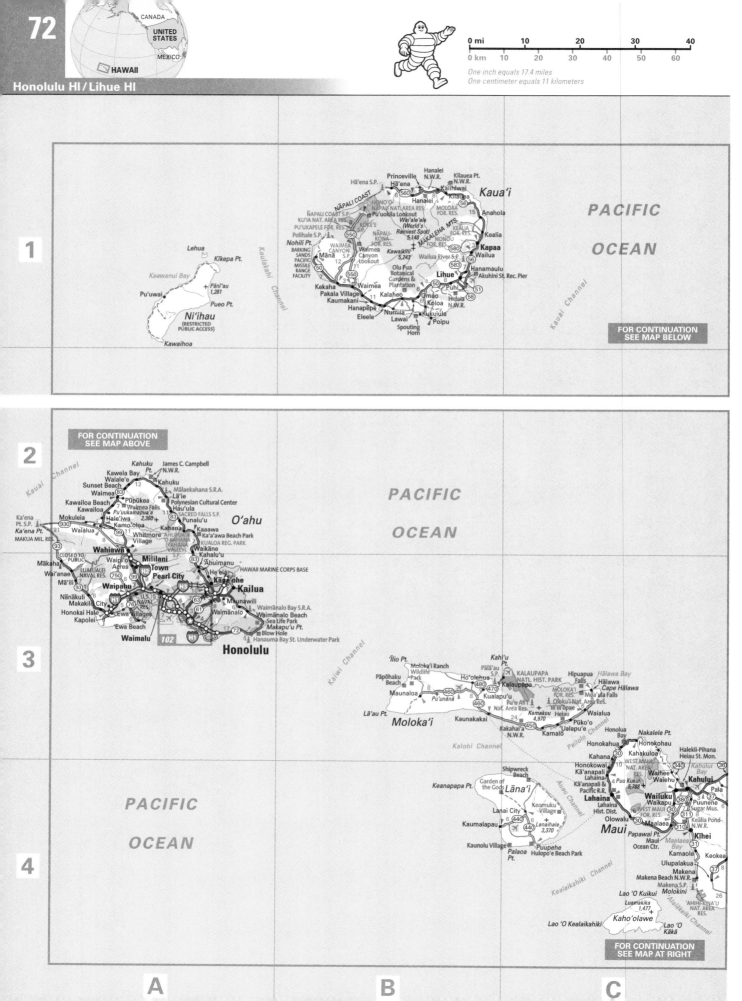

0 mi   10   20   30   40
0 km   10   20   30   40   50   60

One inch equals 17.4 miles
One centimeter equals 11 kilometers

PACIFIC OCEAN

Kaua'i

Hanalei N.W.R.
Kilauea Pt. N.W.R.
Hā'ena S.P.
Princeville
Hā'ena
Hanalei
Kailhiwai
Kilauea
NĀPALI COAST
HONO'O NĀPALI NAT. AREA RES.
Pu'uokila Lookout
Wai'ale'ale (World's Rainiest Spot)
NĀPALI COAST S.P.
KU'IA NAT. AREA RES.
PU'UKAPELE FOR. RES.
Polihale S.P.
Nohili Pt.
BARKING SANDS PACIFIC MISSILE RANGE FACILITY
Lehua
Kīkepa Pt.
Keawanui Bay
Pānī'au 1,281
Pu'uwai
Pueo Pt.
Kawaihoa
Ni'ihau (RESTRICTED PUBLIC ACCESS)

MAKALEHA MTS.
Kawaikīni 5,243
Anahola
Kealia
Kapaa
Wailua
Hanamaulu
Akuhini St. Rec. Pier
Lihue
Puhi
KEALIA FOR. RES.
NONOU FOR. RES.
MOLOA'A FOR. RES.
Wailua River S.P.
Olu Pua Botanical Gardens & Plantation
Māna
Waimea
Kekaha
Pakala Village
Kaumakani
Hanapēpē
Eleele
Kalaheo
Ōmao
Koloa
Numila
Lawai
Kukuiula
Poipu
Spouting Horn
Hulaia N.W.R.
WAIMEA CANYON S.P.
Waimea Canyon Lookout
KOKE'E

Kaulakahi Channel
Kauai Channel

FOR CONTINUATION SEE MAP BELOW

PACIFIC OCEAN

Kahuku Pt.
James C. Campbell N.W.R.
Kawela Bay
Waiale'e
Sunset Beach
Waimea
Kahuku
Mālaekahana S.R.A.
Lā'ie
Polynesian Cultural Center
Hau'ula
SACRED FALLS S.P.
Kawailoa Beach
Kawailoa
Mokuleia
Pūpūkea
Waimea Falls
Hale'iwa
Pu'uakainapua'a 2,360
Kamo'oloa
Waialua
Whitmore Village
Kahana
Punalu'u
Kaaawa
Ka'a'awa Beach Park
AHUPUA'A O KAHANA (KAHANA VALLEY) S.P.
KUALOA REG. PARK
Waikāne
Kahalu'u
Waiahole
Ka'ena Pt. S.P.
Ka'ena Pt.
MAKUA MIL. RES.
Mākaha
Wai'anae
Mā'ili
Nānākuli
Makakilo City
Honokai Hale
Kapolei
'Ewa Villages
'Ewa Beach
Waimalu
LUALUALEI NAVAL RES.
Wahiawā
Mililani Town
Pearl City
Waipi'o Acres
Waipahu
U.S. NAVAL RES.
Ahuimanu
He'eia
Kāne'ohe
HAWAII MARINE CORPS BASE
Kailua
Maunawili
Waimānalo
Waimānalo Bay S.R.A.
Waimānalo Beach
Sea Life Park
Makapu'u Pt.
Blow Hole
Hanauma Bay St. Underwater Park
Honolulu

O'ahu

Kauai Channel
Kaiwi Channel

PACIFIC OCEAN

FOR CONTINUATION SEE MAP ABOVE

'Ilio Pt.
Moloka'i Ranch Wildlife Park
Pāpōhaku Beach
Maunaloa
Lā'au Pt.
Pu'unānā
Ho'olehua
Kahi'u Pt.
Pala'au
KALAUPAPA NATL. HIST. PARK
Kalaupapa
Kualapu'u
MOLOKA'I FOR. RES.
Oloku'i Nat. Area Res.
Nat. Area Res.
Kaunakakai
Kamakou 4,970
'Ili'ili'ōpae Heiau
Kahahai'a N.W.R.
Kamalō
Ualapu'e
Pūko'o
Waialua
Mo'o'ula Falls
Hipuapua Falls
Hālawa
Cape Hālawa
Hālawa Bay

Moloka'i

Kalohi Channel
Pailolo Channel

Nakalele Pt.
Honokohau
Honolua Bay
Honokahua
Kahana
Kahakuloa
Halekii-Pihana Heiau St. Mon.
WEST MAUI NAT. AREA RES.
Honokowai
Kā'anapali
Kā'anapali & Pacific R.R.
Lahaina
Lahaina Hist. Dist.
Puu Kukui 5,788
Waihe'e
Waiehu
Kahului
Paia
Wailuku
Waikapu
Puunene
Maui Sugar Mus.
Olowalu
Māalaea
Papawai Pt.
Māalaea Bay
Maui Ocean Ctr.
Kīhei
Kamaole
Keokea
Ulupalakua
Makena
Makena Beach N.W.R.
Makena S.P.
Kahului Bay

Maui

Shipwreck Beach
Keanapapa Pt.
Garden of the Gods
Lāna'i
Keomuku Village
Lanai City
Kaumalapau
Kaunolu Village
Palaoa Pt.
Lanaihale 3,370
Puupehe
Hulopo'e Beach Park

'Au'au Channel

PACIFIC OCEAN

Lao 'O Kuikui
Luamakika 1,477
Lao 'O Kealaikahiki
Kaho'olawe
Lao 'O Kākā

'AHIHI-KINA'U NAT. AREA RES.
Molokini

Kealaikahiki Channel
Alalākeiki Channel

FOR CONTINUATION SEE MAP AT RIGHT

A   B   C

UNITED STATES
CANADA
MEXICO
HAWAII

SEE ALSO DISTANCE AND DRIVING TIME MAP ON PAGE 126

DRIVING DISTANCES IN MILES	HANA	HILO	HONOLULU	HOOLEHUA	KAHULUI	KAILUA	KAILUA-KONA	LAHAINA	LANAI CITY	LIHUE	WAHIAWA	WAIMEA
HILO	149*		217*	169*	121*	235*	88	142*	155*	319*	234*	54
HONOLULU	129*	217*		54*	101*	14	185*	92*	74*	102*	23	172*
KAHULUI	42	121*	101*	76		109*	57	202*	118*	79*		
LIHUE	230*	319*	102*	156	202*	120*	285*	225*	176*		119	174*

*DISTANCE INCLUDES AIR TRAVEL

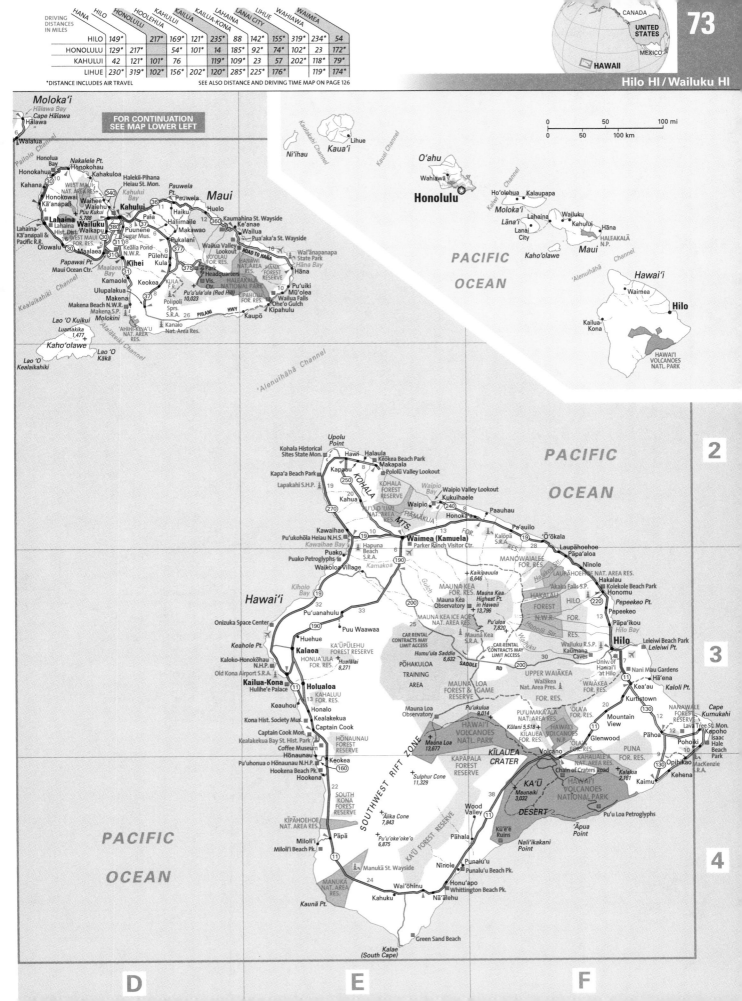

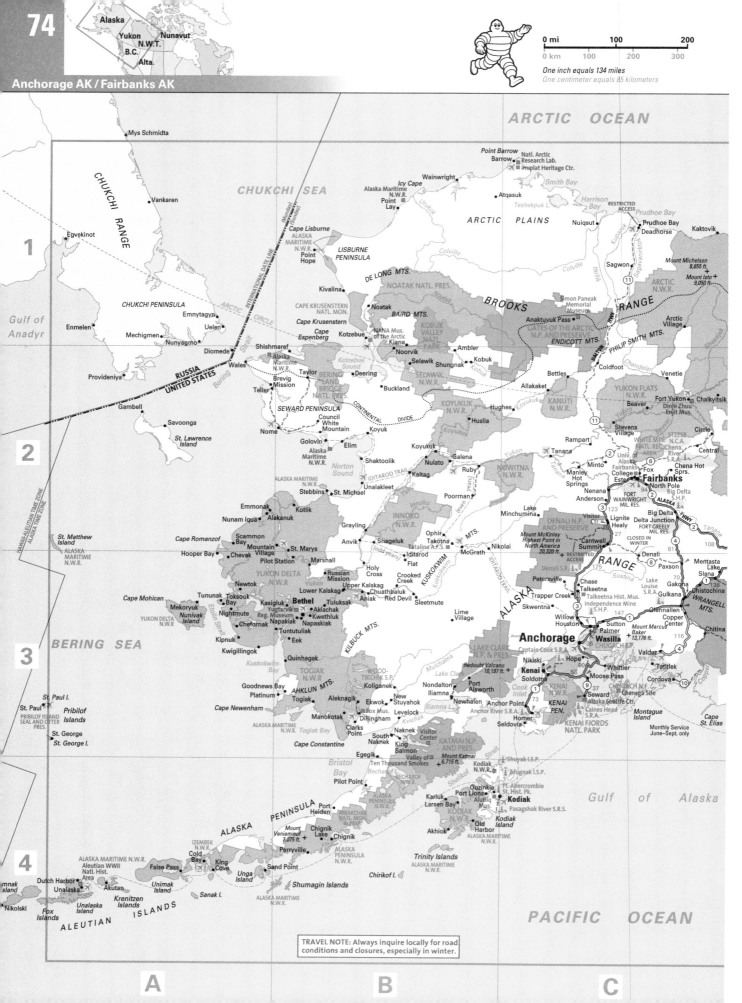

DRIVING DISTANCES IN MILES	ANCHORAGE, AK	DAWSON CREEK, BC	DENALI NP, AK	FAIRBANKS, AK	HOMER, AK	JUNEAU, AK	PRINCE GEORGE, BC	PRINCE RUPERT, BC	SKAGWAY, AK	TOK, AK	WHITEHORSE, YT	YELLOWKNIFE, NT
ANCHORAGE, AK		1516	275	378	225	841*	1679	1514	807	323	697	1844
DAWSON CREEK, BC	1516		1503	1400	1740	963*	224	625	862	1193	819	741
FAIRBANKS, AK	378	1400	103		603	726*	1564	1398	691	207	581	1729
WHITEHORSE, YT	697	819	684	581		921	211*	982	817	110	374	1147

*DISTANCE INCLUDES FERRY TRAVEL

SEE ALSO DISTANCE AND DRIVING TIME MAP ON PAGE 126

Distances in the U.S. shown in miles.
Distances in Canada shown in kilometers.

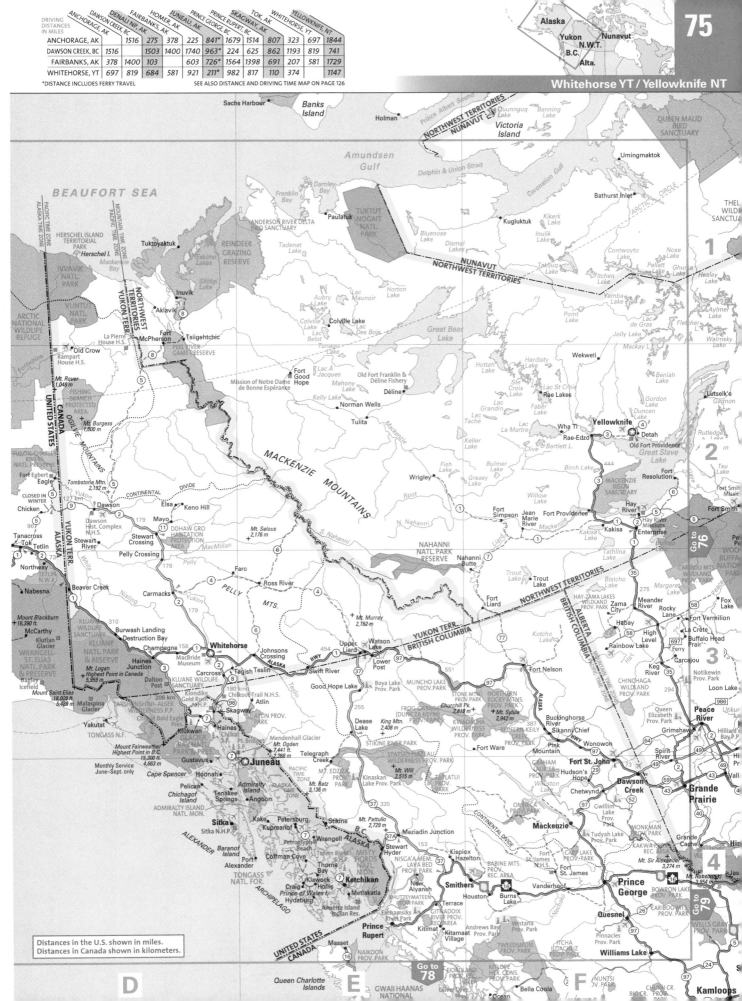

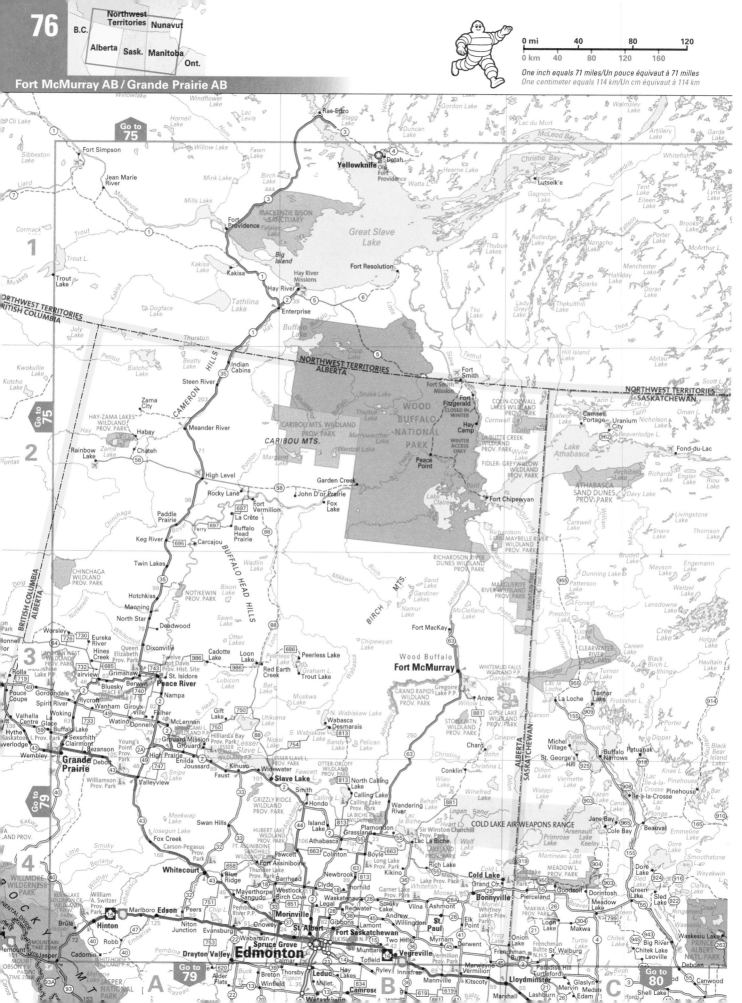

76

Northwest Territories   Nunavut
B.C.
Alberta   Sask.   Manitoba
Ont.

Fort McMurray AB / Grande Prairie AB

One inch equals 71 miles/Un pouce équivaut à 71 milles
One centimeter equals 114 km/Un cm équivaut à 114 km

0 mi   40   80   120
0 km   40   80   120   160

DRIVING DISTANCES IN KM / DISTANCES ROUTIÈRES EN KM

	EDMONTON, AB	FLIN FLON, MB	FORT MCMURRAY, AB	GRANDE PRAIRIE, AB	GRAND RAPIDS, MB	JASPER, AB	LA RONGE, SK	PEACE RIVER, AB	PRINCE ALBERT, SK	SLAVE LAKE, AB	THOMPSON, MB	YELLOWKNIFE, NT
EDMONTON, AB		955	439	462	1172	367	830	484	577	251	1302	1405
JASPER, AB	367	1312	796	397	1530		1187	578	932	464	1657	1532
PRINCE ALBERT, SK	577	375	944	1035	604	932	238	1055		822	729	2065
THOMPSON, MB	1302	380	1668	1760	328	1657	704	1779	729	1547		2790

SEE ALSO DISTANCE AND DRIVING TIME MAP ON PAGE 126 / VOIR AUSSI CARTE DES DISTANCES ET DES TEMPS DE PARCOURS PAGE 126

DISTANCES IN CANADA SHOWN IN KILOMETERS

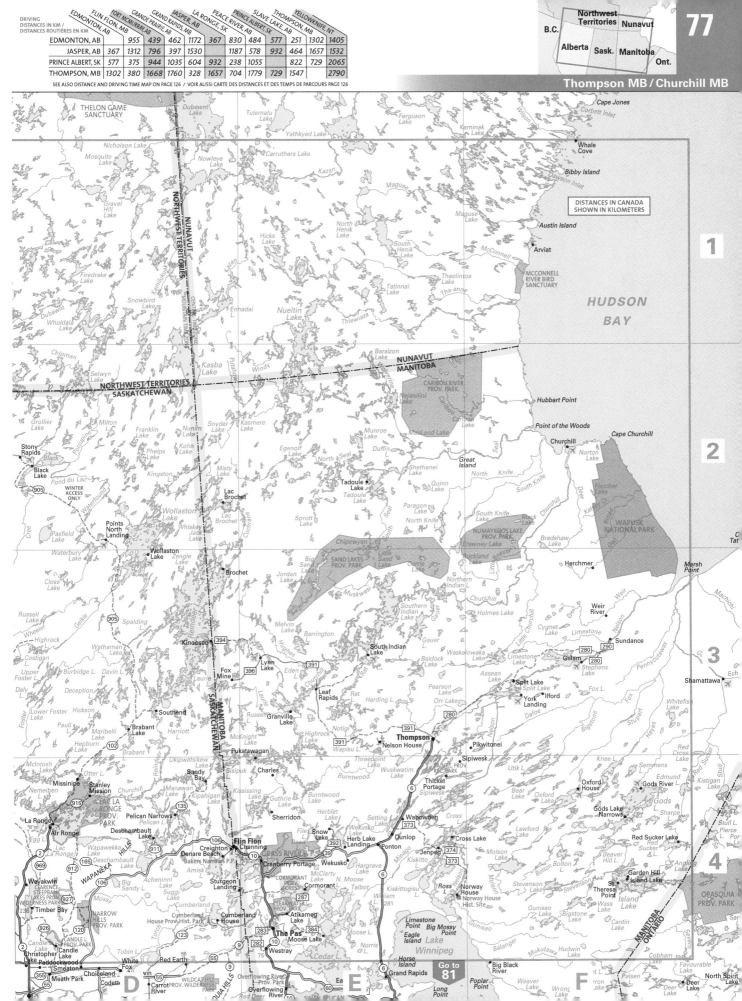

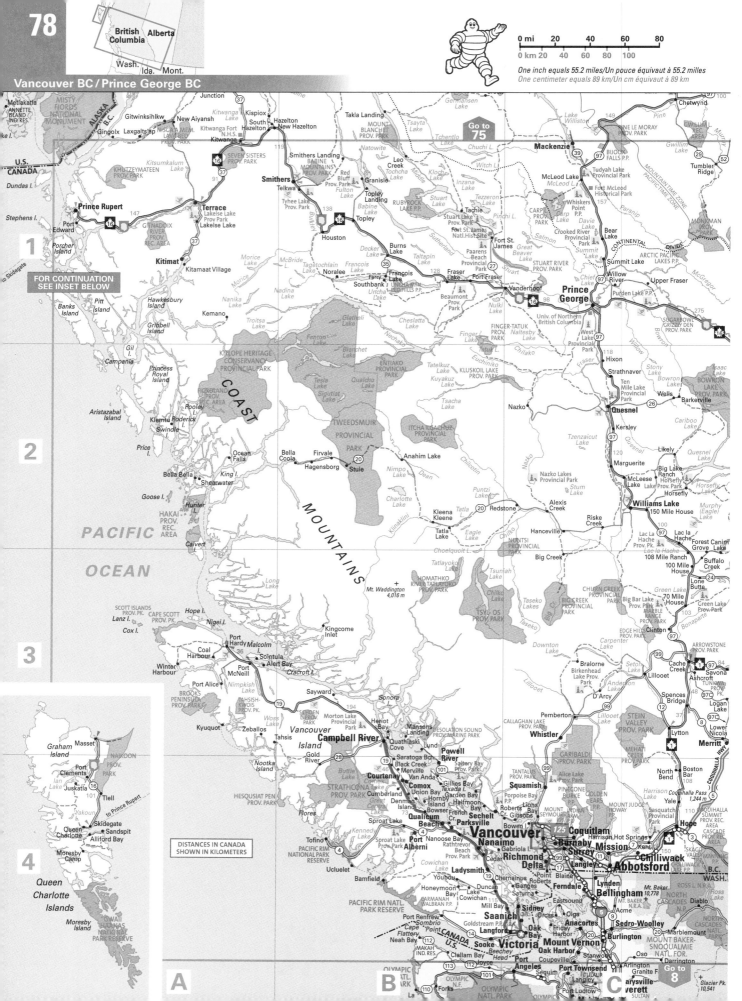

British Columbia · Alberta
Wash. · Ida. · Mont.

0 mi 20 40 60 80
0 km 20 40 60 80 100
One inch equals 55.2 miles/Un pouce équivaut à 55.2 milles
One centimeter equals 89 km/Un cm équivaut à 89 km

FOR CONTINUATION
SEE INSET BELOW

PACIFIC

OCEAN

COAST

MOUNTAINS

Go to 75

DISTANCES IN CANADA
SHOWN IN KILOMETERS

Queen
Charlotte
Islands

Go to 8

British Columbia   Alberta

Wash.   Ida.   Mont.

DRIVING DISTANCES IN KM / DISTANCES ROUTIÈRES EN KM

	CALGARY, AB	EDMONTON, AB	GRANDE PRAIRIE, AB	JASPER, AB	KALISPELL, MT	KAMLOOPS, BC	KELOWNA, BC	LETHBRIDGE, AB	PRINCE GEORGE, BC	PRINCE RUPERT, BC	VANCOUVER, BC	VICTORIA, BC
CALGARY, AB		296	750	396	427	620	602	216	789	1513	975	1013*
EDMONTON, AB	296		462	367	723	800	897	512	737	1461	1155	1193*
PRINCE GEORGE, BC	789	737	492	376	1050	525	688	978		724	778	821*
VANCOUVER, BC	975	1155	1134	784	1044	340	378	1124	778	1502		93*

*DISTANCE INCLUDES FERRY TRAVEL / LA DISTANCE INCLUT LE VOYAGE EN TRAVERSIER

SEE ALSO DISTANCE AND DRIVING TIME MAP ON PAGE 126 / VOIR AUSSI CARTE DES DISTANCES ET DES TEMPS DE PARCOURS PAGE 126

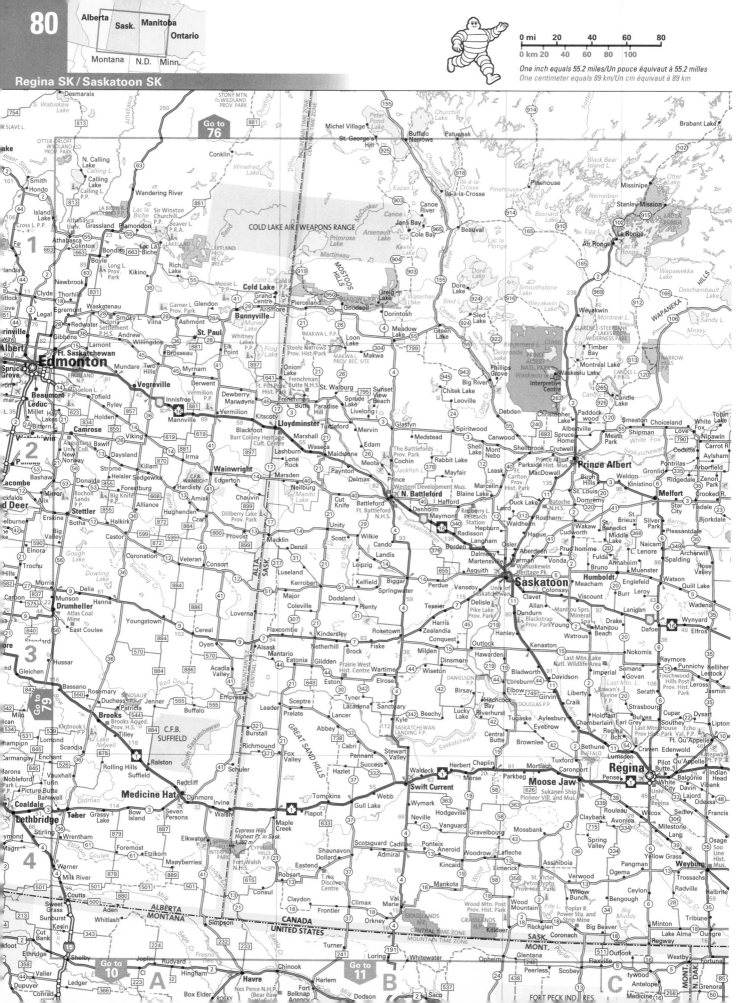

0 mi 20 40 60 80
0 km 20 40 60 80 100

One inch equals 55.2 miles/Un pouce équivaut à 55.2 milles
One centimeter equals 89 km/Un cm équivaut à 89 km

81

DRIVING DISTANCES IN KM /
DISTANCES ROUTIÈRES EN KM

	BRANDON, MB	EDMONTON, AB	FLIN FLON, MB	GRAND RAPIDS, MB	LETHBRIDGE, AB	PRINCE ALBERT, SK	REGINA, SK	SASKATOON, SK	SWAN RIVER, MB	SWIFT CURRENT, SK	THOMPSON, MB	WINNIPEG, MB
BRANDON, MB		1163	676	525	1002	670	377	639	333	598	855	216
EDMONTON, AB	1163		955	1172	512	577	779	527	930	682	1302	1358
SASKATOON, SK	639	527	508	654	652	141	261		409	267	821	837
WINNIPEG, MB	216	1358	757	430	1198	816	583	837	489	811	769	

SEE ALSO DISTANCE AND DRIVING TIME MAP ON PAGE 126 / VOIR AUSSI CARTE DES DISTANCES ET DES TEMPS DE PARCOURS PAGE 126

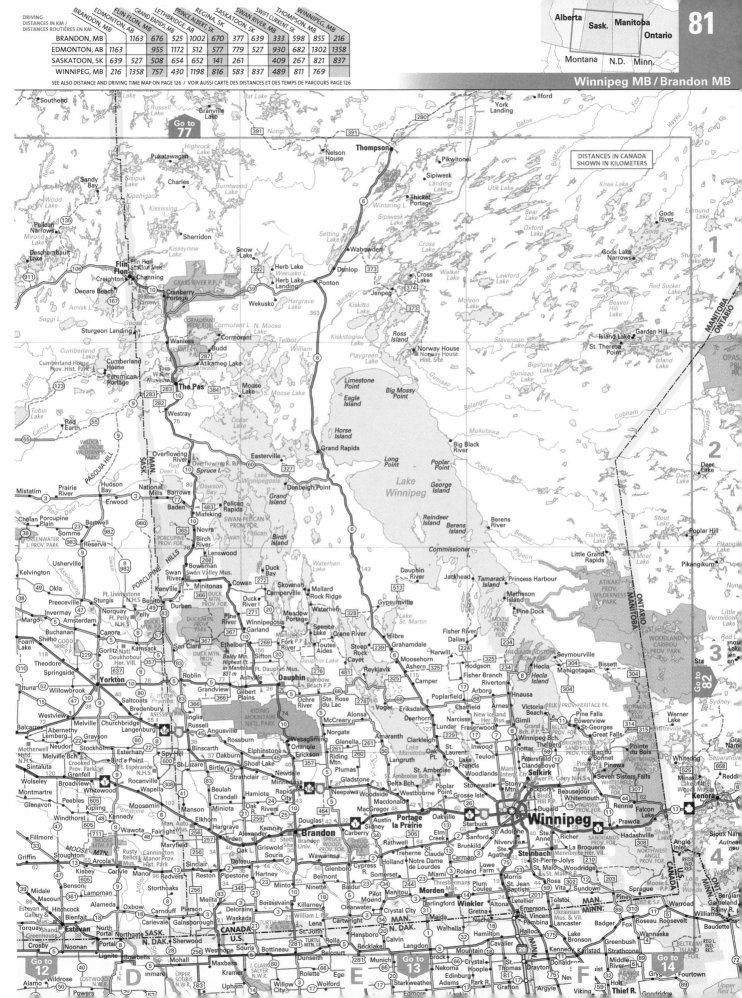

DISTANCES IN CANADA
SHOWN IN KILOMETERS

Go to
77

Go to
82

Go to
12

Go to
13

Go to
14

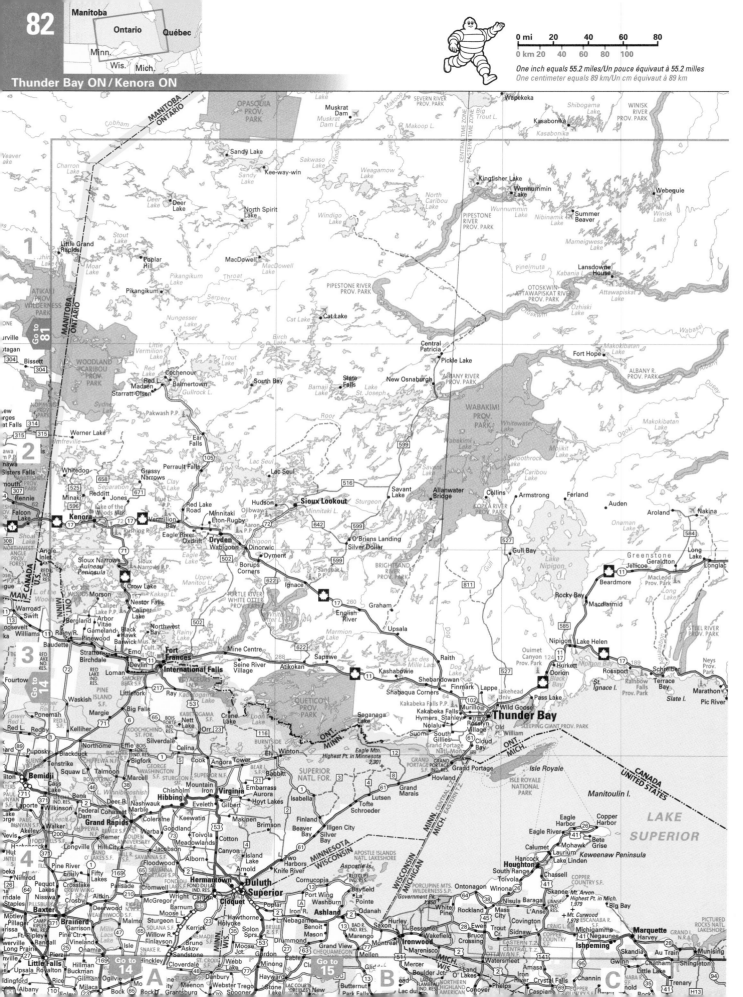

0 mi 20 40 60 80
0 km 20 40 60 80 100

One inch equals 55.2 miles/Un pouce équivaut à 55.2 milles
One centimeter equals 89 km/Un cm équivaut à 89 km

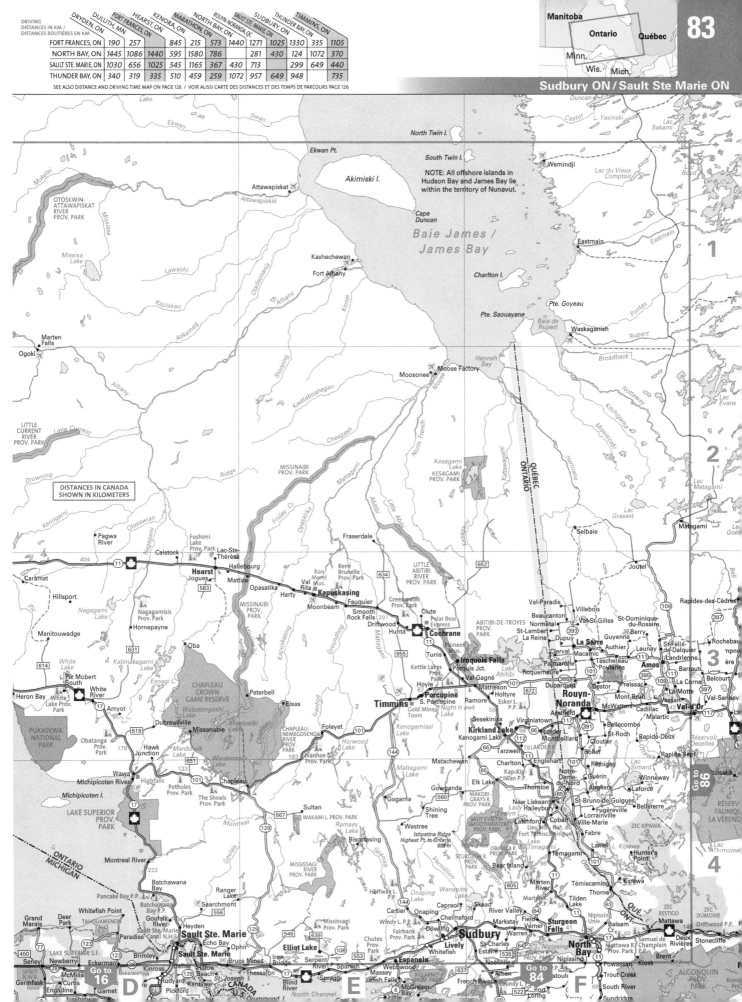

DRIVING DISTANCES IN KM / DISTANCES ROUTIÈRES EN KM

	DULUTH, MN	DRYDEN, ON	FORT FRANCES, ON	HEARST, ON	KENORA, ON	MARATHON, ON	NORTH BAY, ON	ROUYN-NORANDA, QC	SAULT STE. MARIE, ON	SUDBURY, ON	THUNDER BAY, ON	TIMMINS, ON
FORT FRANCES, ON	190	257		845	215	573	1440	1271	1025	1330	335	1105
NORTH BAY, ON	1445	1086	1440	595	1580	786		281	430	124	1072	370
SAULT STE. MARIE, ON	1030	656	1025	545	1165	367	430	713		299	649	440
THUNDER BAY, ON	340	319	335	510	459	259	1072	957	649	948		735

SEE ALSO DISTANCE AND DRIVING TIME MAP ON PAGE 126 / VOIR AUSSI CARTE DES DISTANCES ET DES TEMPS DE PARCOURS PAGE 126

DISTANCES IN CANADA SHOWN IN KILOMETERS

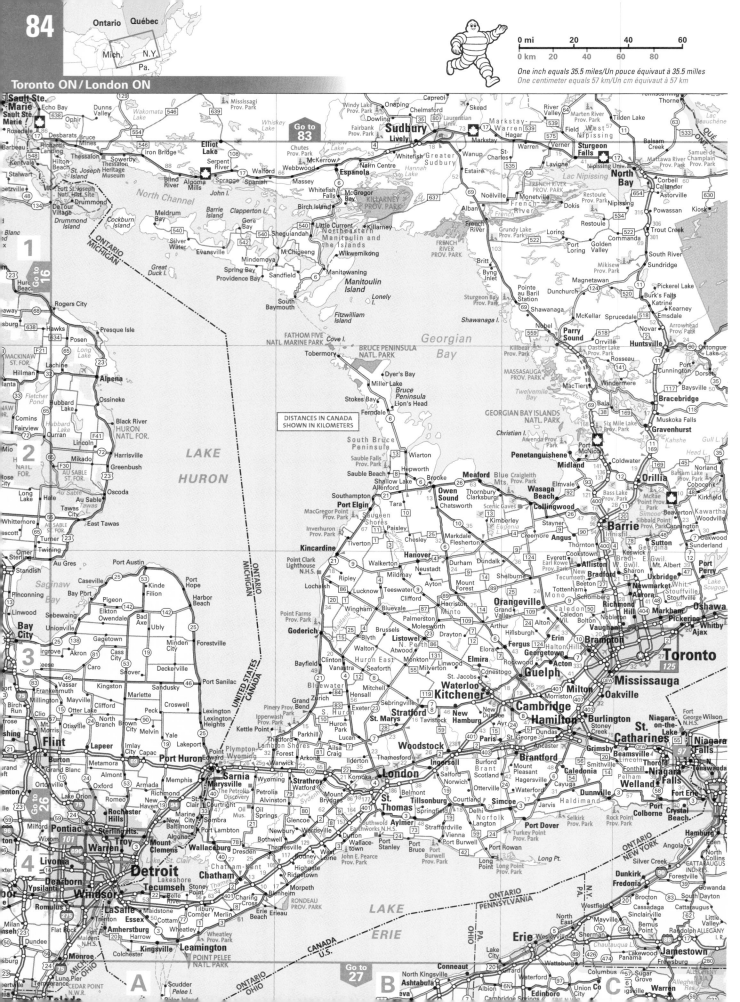

Ontario Québec

Mich. N.Y.

Pa.

0 mi   20   40   60
0 km 20 40 60 80
One inch equals 35.5 miles/Un pouce équivaut à 35.5 miles
One centimeter equals 57 km/Un cm équivaut à 57 km

DISTANCES IN CANADA
SHOWN IN KILOMETERS

LAKE HURON

Georgian Bay

Manitoulin Island

LAKE ERIE

ONTARIO
MICHIGAN

UNITED STATES
CANADA

ONTARIO
NEW YORK

ONTARIO
PENNSYLVANIA

ONTARIO
OHIO

CANADA
U.S.

DRIVING DISTANCES IN KM /
DISTANCES ROUTIÈRES EN KM

	BARRIE, ON	DETROIT, MI	KINGSTON, ON	LONDON, ON	MONTRÉAL, QC	NIAGARA FALLS, ON	NORTH BAY, ON	OTTAWA, ON	SAULT STE. MARIE, ON	SUDBURY, ON	SYRACUSE, NY	TORONTO, ON
DETROIT, MI	436		613	212	899	389	663	807	563	741	692	373
KINGSTON, ON	330	613		434	290	390	460	179	875	600	211	251
MONTRÉAL, QC	610	899	290	715		670	544	194	967	685	404	531
TORONTO, ON	105	373	251	183	531	145	336	431	674	407	393	

SEE ALSO DISTANCE AND DRIVING TIME MAP ON PAGE 126 / VOIR AUSSI CARTE DES DISTANCES ET DES TEMPS DE PARCOURS PAGE 126

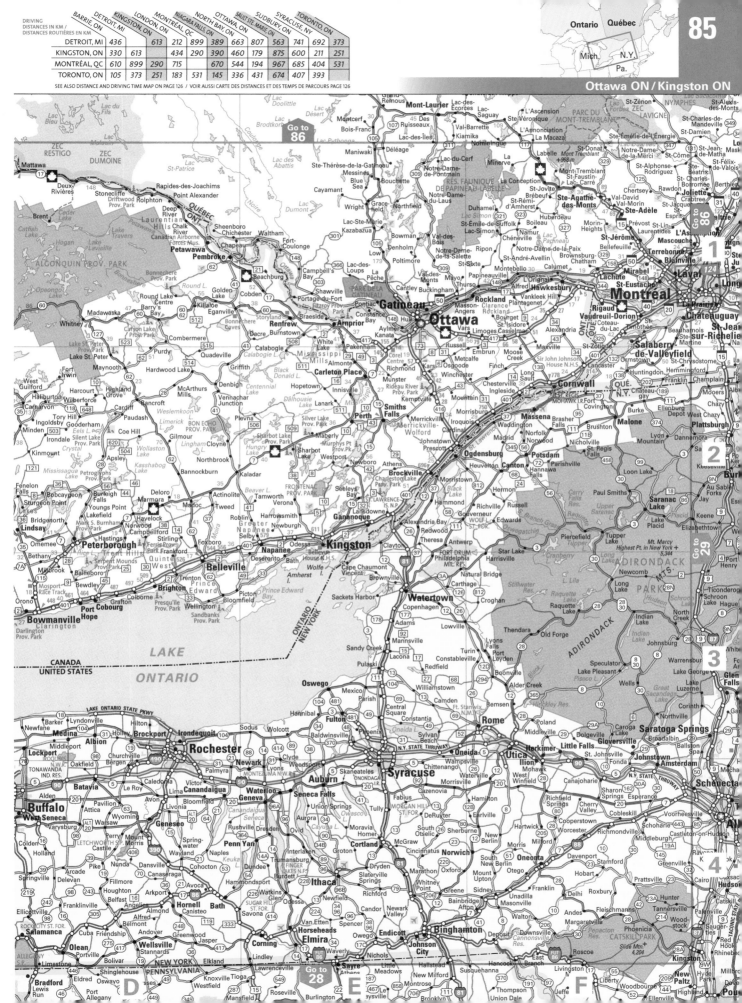

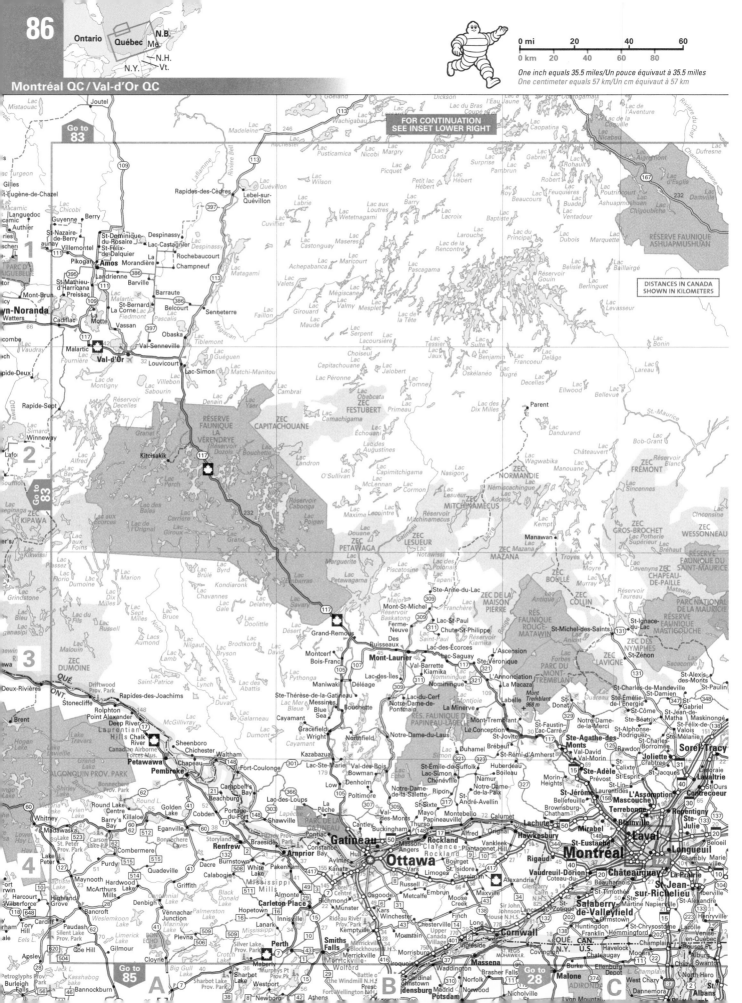

Ontario Québec N.B. Me.
N.H.
N.Y. Vt.

0 mi   20   40   60
0 km 20 40 60 80

One inch equals 35.5 miles/Un pouce équivaut à 35.5 milles
One centimeter equals 57 km/Un cm équivaut à 57 km

FOR CONTINUATION
SEE INSET LOWER RIGHT

DISTANCES IN CANADA
SHOWN IN KILOMETERS

Go to 83
Go to 83
Go to 85
Go to 28

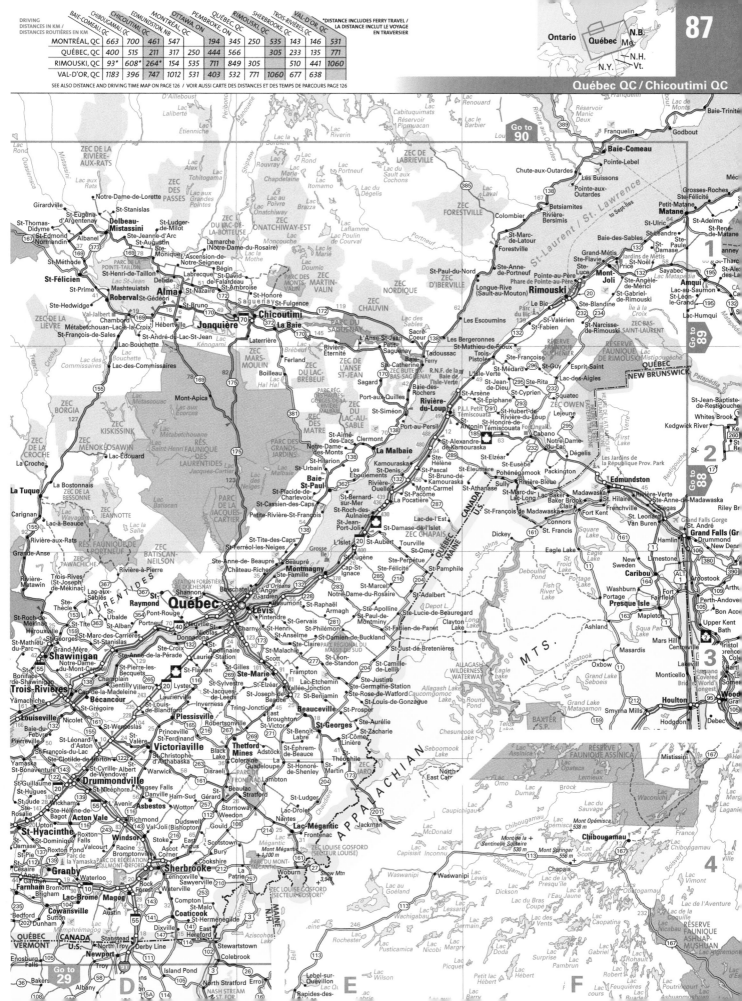

DRIVING DISTANCES IN KM / DISTANCES ROUTIÈRES EN KM

*DISTANCE INCLUDES FERRY TRAVEL / LA DISTANCE INCLUT LE VOYAGE EN TRAVERSIER

	BAIE-COMEAU, QC	CHIBOUGAMAU, QC	CHICOUTIMI, QC	EDMUNDSTON, NB	MONTRÉAL, QC	OTTAWA, ON	PEMBROKE, ON	QUÉBEC, QC	RIMOUSKI, QC	SHERBROOKE, QC	TROIS-RIVIÈRES, QC	VAL-D'OR, QC
MONTRÉAL, QC	663	700	461	547		194	345	250	535	143	146	531
QUÉBEC, QC	400	515	211	317	250	444	566		305	233	135	771
RIMOUSKI, QC	93*	608*	264*	154	535	711	849	305		510	441	1060
VAL-D'OR, QC	1183	396	747	1012	531	403	532	771	1060	677	638	

SEE ALSO DISTANCE AND DRIVING TIME MAP ON PAGE 126 / VOIR AUSSI CARTE DES DISTANCES ET DES TEMPS DE PARCOURS PAGE 126

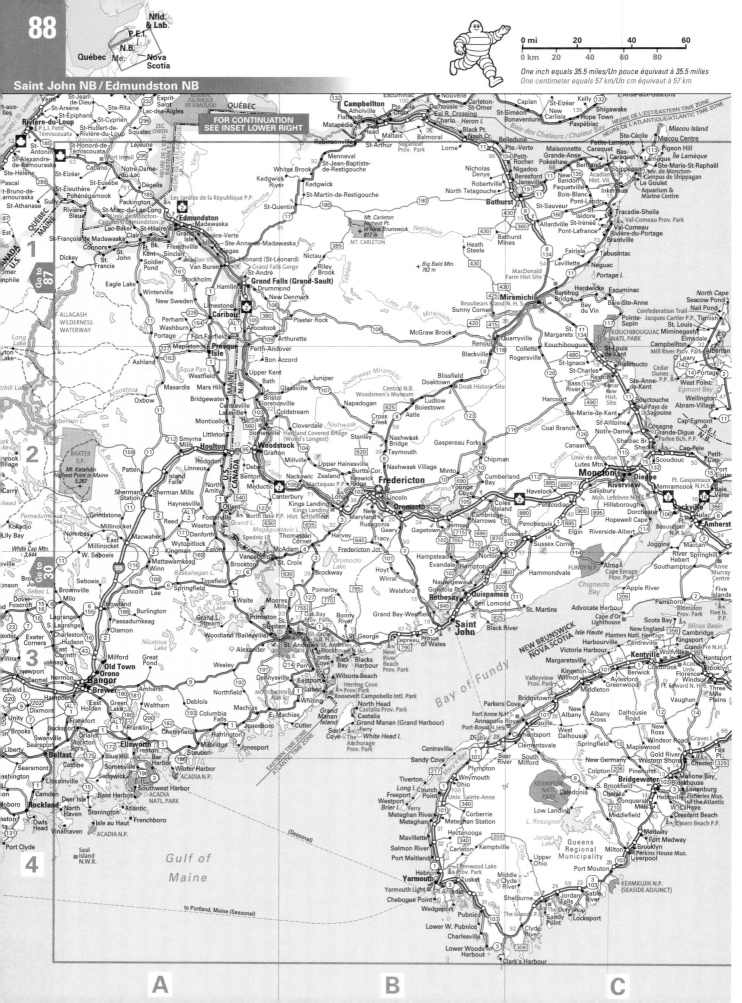

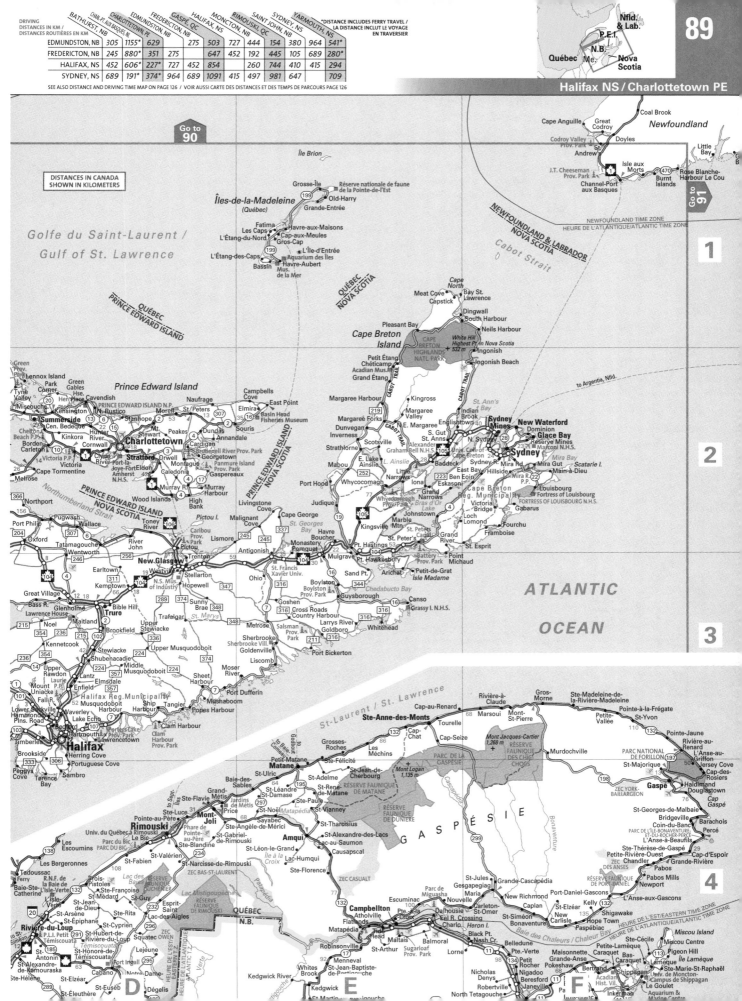

DRIVING DISTANCES IN KM / DISTANCES ROUTIÈRES EN KM

	BATHURST, NB	CHAN.-PT AUX BASQUES, NL	CHARLOTTETOWN, PE	EDMUNDSTON, NB	FREDERICTON, NB	GASPÉ, QC	HALIFAX, NS	MONCTON, NB	RIMOUSKI, QC	SAINT JOHN, NB	SYDNEY, NS	YARMOUTH, NS
EDMUNDSTON, NB	305	1155*	629		275	503	727	444	154	380	964	541*
FREDERICTON, NB	245	880*	351	275		647	452	192	445	105	689	280*
HALIFAX, NS	452	606*	227*	727	452	854		260	744	410	415	294
SYDNEY, NS	689	191*	374*	964	689	1091	415	497	981	647		709

*DISTANCE INCLUDES FERRY TRAVEL / LA DISTANCE INCLUT LE VOYAGE EN TRAVERSIER

SEE ALSO DISTANCE AND DRIVING TIME MAP ON PAGE 126 / VOIR AUSSI CARTE DES DISTANCES ET DES TEMPS DE PARCOURS PAGE 126

DISTANCES IN CANADA SHOWN IN KILOMETERS

DRIVING DISTANCES IN KM / DISTANCES ROUTIÈRES EN KM

	ARGENTIA, NL	CHAN.-PT. AUX BASQUES, NL	DEER LAKE, NL	GANDER, NL	ST. JOHN'S, NL	SYDNEY, NS
ARGENTIA, NL		845	588	291	134	457*
CHAN-PT. AUX BASQUES, NL	845		257	554	875	191*
ST. JOHN'S, NL	134	875	618	321		591*
SYDNEY, NS	457*	191*	448*	745*	591*	

*DISTANCE INCLUDES FERRY TRAVEL / LA DISTANCE INCLUT LE VOYAGE EN TRAVERSIER

SEE ALSO DISTANCE AND DRIVING TIME MAP ON PAGE 126 / VOIR AUSSI CARTE DES DISTANCES ET DES TEMPS DE PARCOURS PAGE 126

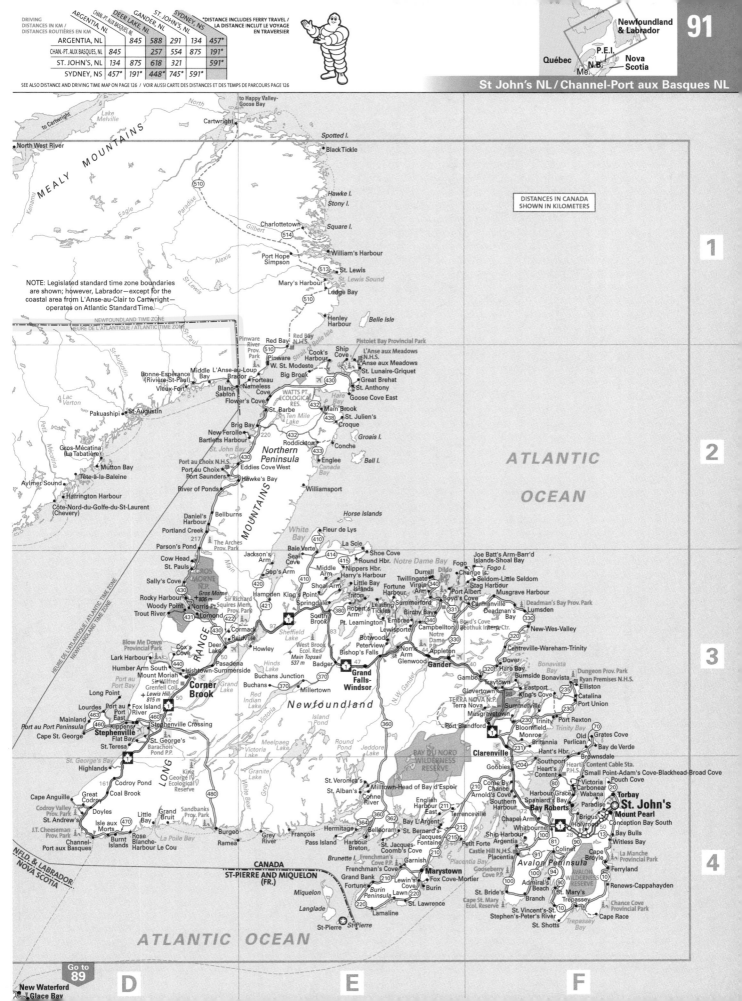

DISTANCES IN CANADA SHOWN IN KILOMETERS

NOTE: Legislated standard time zone boundaries are shown; however, Labrador—except for the coastal area from L'Anse-au-Clair to Cartwright—operates on Atlantic Standard Time.

Go to 89

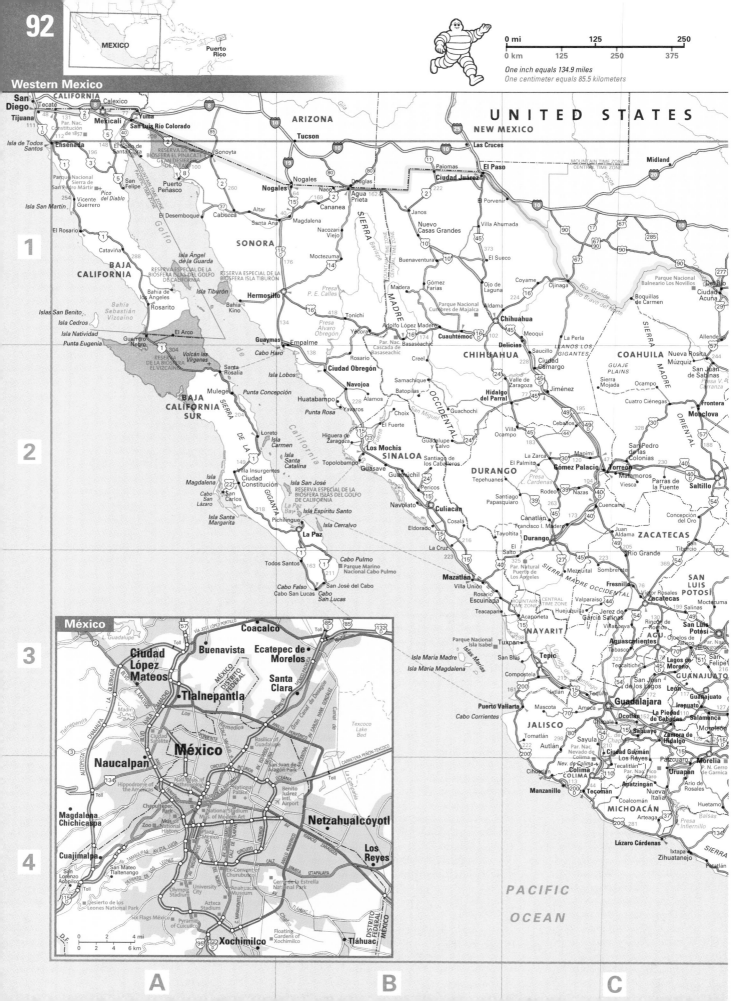

DRIVING DISTANCES IN KILOMETERS

*DISTANCE INCLUDES FERRY TRAVEL

	ACAPULCO	CANCÚN	CHIHUAHUA	GUADALAJARA	LA PAZ	MAZATLÁN	MÉXICO	MONTERREY	PUEBLA	SAN LUIS POTOSÍ	TIJUANA	TUXTLA GUTIÉRREZ
CHIHUAHUA	1913	3290		1283	1432*	896	1538	808	1651	1155	1456	2417
GUADALAJARA	897	2275	1283		959*	523	578	758	691	336	2121	1510
MÉXICO	422	1736	1538	578	1537*	1081		892	133	381	2700	932
MONTERREY	1404	2416	808	758	1357*	901	892		1035	509	2362	1609

SEE ALSO DISTANCE AND DRIVING TIME MAP ON PAGE 126

MEXICO

Acámbaro.....D3
Acaponeta.....C3
Acapulco.....D4
Acayucan.....E4
Agua Prieta.....B1
Aguascalientes.....C3
Álamo.....D3
Aldama.....B1
Allende.....C1
Alvarado.....E4
Anáhuac.....D2
Apatzingán.....C4
Arriaga.....E4
Atoyac.....D4
Autlán.....C4
Caborca.....A1
Cabo San Lucas.....B3
Campeche.....F3
Cananea.....B1

Cancún.....F3
Cárdenas.....E4
Celaya.....C3
Cerro Azul.....D3
Champotón.....F3
Chapala.....C3
Chetumal.....F4
Chihuahua.....B1
Chilpancingo.....D4
Cholula.....D4
Cihuatlán.....C4
Cintalapa.....E4
Cd. Acuña.....C1
Cd. Altamirano.....D4
Cd. Camargo.....C2
Cd. Constitución.....A2
Cd. del Carmen.....E4
Cd. Hidalgo.....D3
Cd. Juárez.....B1

Cd. Madero.....D3
Cd. Mante.....D3
Cd. Miguel Alemán.....D2
Cd. Obregón.....B2
Cd. Valles.....D3
Cd. Victoria.....D3
Coatzacoalcos.....E4
Colima.....C4
Comitán.....E4
Compostela.....C3
Córdoba.....D4
Cosamaloapan.....E4
Cozumel.....F3
Cuauhtémoc.....B1
Cuautla.....D4
Cuernavaca.....D4
Culiacán.....B2
Delicias.....C1
Durango.....C2
Ébano.....D3
Emiliano Zapata.....E4
Empalme.....B1
Ensenada.....A1
Escárcega.....F4
Escuinapa.....B3
Felipe Carrillo Puerto.....F3
Fresnillo.....C3
Frontera.....E4
Frontera.....E4
Gómez Palacio.....C2

Guadalajara.....C3
Guamúchil.....B2
Guanajuato.....C3
Guasave.....B2
Guaymas.....B1
Hermosillo.....B1
Hidalgo del Parral.....C2
Huajuapan de León.....D4
Huatabampo.....B2
Huejutla.....D3
Huetamo.....C4
Huixtla.....E4
Iguala.....D4
Irapuato.....C3
Ixmiquilpan.....D3
Ixtlán.....D4
Jerez de García Salinas.....C3
Jiménez.....C2
Juchitán.....E4
Lagos de Moreno.....C3
La Paz.....B2
La Piedad de Cabadas.....C3
Lázaro Cárdenas.....C4
León.....C3
Linares.....D2
Loreto.....A2
Los Mochis.....B2
Los Reyes.....C4

Macuspana.....E4
Magdalena.....B1
Manzanillo.....C4
Matamoros.....D2
Matamoros.....D2
Matamoros.....D3
Matehuala.....D3
Matías Romero.....E4
Mazatlán.....B3
Meoqui.....C1
Mérida.....F3
Mexicali.....A1
México.....D4
Miahuatlán.....D4
Minatitlán.....E4
Monclova.....C2
Montemorelos.....D2
Monterrey.....D2
Morelia.....C3
Moroleón.....C3
Múzquiz.....C2
Navojoa.....B2
Navolato.....B2
Nogales.....B1
Nueva Italia.....C4
Nueva Rosita.....C2
Nuevo Casas Grandes.....B1
Nuevo Laredo.....D2
Oaxaca.....D4
Ocotlán.....C3

Orizaba.....D4
Pachuca.....D3
Papantla.....D3
Paraíso.....E4
Parras de la Fuente.....C2
Pátzcuaro.....C4
Perote.....D3
Petatlán.....C4
Piedras Negras.....D1
Playa del Carmen.....F3
Poza Rica.....D3
Progreso.....F3
Puebla.....D4
Puerto Escondido.....D4
Puerto Peñasco.....A1
Puerto Vallarta.....C3
Querétaro.....D3
Reynosa.....D2
Rincón de Romos.....C3
Río Bravo.....D2
Río Grande.....C3
Río Verde.....D3
Sabinas Hidalgo.....D2
Sahuayo.....C3
Salamanca.....C3
Salina Cruz.....E4
Saltillo.....C2
San Andrés Tuxtla.....E4
San Cristóbal de las Casas.....E4

San Felipe.....C3
San Fernando.....D2
San José del Cabo.....B3
San Juan de los Lagos.....C3
San Juan del Río.....D3
San Luis de la Paz.....C3
San Luis Potosí.....C3
San Luis Río Colorado.....A1
San Miguel de Allende.....C3
San Pedro de las Colonias.....C2
Santiago Papasquiaro.....C2
Sayula.....C3
Sombrerete.....C3
Tamazunchale.....D3
Tampico.....D3
Tantoyuca.....D3
Tapachula.....E4
Taxco.....D4
Teapa.....E4
Tecate.....A1
Tecomán.....C4
Tecpan.....D4
Tehuacán.....D4
Tehuantepec.....E4
Tenancingo.....D4
Tenosique.....E4
Teocaltiche.....C3
Tepic.....C3

Tequila.....C3
Teziutlán.....D3
Ticul.....F3
Tijuana.....A1
Tizayuca.....D3
Tizimín.....F3
Tlapa.....D4
Tlapacoyan.....D3
Tlaxcala.....D4
Toluca.....D4
Tonalá.....E4
Torreón.....C2
Tulancingo.....D3
Tuxpan.....D3
Tuxpan.....C3
Tuxtepec.....D4
Tuxtla Gutiérrez.....E4
Umán.....F3
Uruapan.....C4
Valladolid.....F3
Valle Hermoso.....D2
Veracruz.....D4
Víctor Rosales.....C3
Villahermosa.....E4
Xalapa.....D4
Zacatecas.....C3
Zacatlán.....D3
Zamora de Hidalgo.....C3
Zihuatanejo.....C4
Zitácuaro.....D4

PUERTO RICO

Adjuntas.....E2
Aibonito.....E2
Arecibo.....E2
Bayamón.....E2
Cabo Rojo.....E2
Caguas.....E2
Carolina.....F2
Cayey.....E2
Ceiba.....E2
Coamo.....E2
Culebra.....F2
Fajardo.....F2
Guayama.....E2
Guaynabo.....E2
Humacao.....F2
Isabela.....E2
Lares.....E2
Luquillo.....F2
Manatí.....E2
Mayagüez.....E2
Ponce.....E2
Salinas.....E2
San Germán.....E2
San Juan.....F2
San Sebastián.....E2
Trujillo Alto.....F2
Vega Baja.....F2
Vieques.....F2
Yabucoa.....F2

UNITED STATES

A
Abbeville AL 65 E2
Abbeville GA 65 F2
Abbeville LA 63 E4
Abbeville SC 54 A4
Abbotsford WI 25 D1
Aberdeen ID 19 F3

Albany MO 38 B1
Albany NY 70 A1
Albany OR 17 E1
Albany TX 59 E1
Albemarle NC 54 C2
Albert Lea MN 24 B2
Albertville AL 53 D3
Albertville MN 14 C4

Amboy WA 8 B4
Ambridge PA 41 F1
Amelia LA 63 F4
American Falls ID 19 F3
American Fork UT 34 A1
Americus GA 65 E2
Amery WI 15 D4
Ames IA 24 B4
Amesbury MA 71 E1

Amherst MA 70 C2
Amherst NH 71 D1
Amherst VA 42 A4
Amherstdale WV 41 E4
Amite LA 63 F3
Amity OR 8 A4
Ammon ID 20 A2
Amory MS 52 B4
Amsterdam MT 10 C4

Amsterdam NY 70 A1
Anaconda MT 10 B3
Anacortes WA 8 B1
Anadarko OK 49 F3
Anaheim CA 44 C4
Anahola HI 72 B1
Anahuac TX 61 F1
Anamosa IA 24 C4
Anchorage AK 74 C3
Anchor Pt. AK 74 C3
Andalusia AL 65 D2
Anderson CA 31 E4
Anderson IN 40 C2
Anderson MO 51 D1
Anderson SC 54 A3
Andover KS 37 F4
Andover MA 71 E1
Andrews SC 54 C4
Andrews TX 58 C1
Angel Fire NM 48 A1
Angels Camp CA 32 A4
Angier NC 55 D2
Angleton TX 61 F1
Angola IN 26 B4
Angoon AK 75 E4
Angwin CA 31 E3
Aniak AK 74 B3
Ankeny IA 24 B4
Anna IL 39 F4
Annandale MN 14 B4
Annandale VA 124 A3
Annapolis MD 68 B4
Ann Arbor MI 26 C3
Anniston AL 53 D4
Annville PA 68 B2
Anoka MN 14 C4
Anson TX 59 E1
Ansonia CT 70 B3
Ansted WV 41 F4
Antelope SD 22 C2
Anthony KS 37 E4
Anthony NM 57 E2

Anthony TX 57 E2
Antigo WI 15 F4
Antioch CA 31 F4
Antioch IL 25 E3
Antlers OK 50 C3
Apache OK 49 F3
Apache Jct. AZ 46 B4
Apex NC 55 D2
Apopka FL 67 E1
Appalachia VA 54 A1
Appleton MN 14 A4
Appleton WI 25 E1
Apple Valley CA 45 D3
Apple Valley MN 24 B1
Appomattox VA 42 B4
Aptos CA 31 D4
Arab AL 53 D3
Aransas Pass TX 61 E2
Arapahoe NE 37 D1
Arapahoe WY 20 C3
Arbutus MD 95 A3
Arcade CA 119 D1
Arcadia CA 106 C2
Arcadia FL 67 E2
Arcadia LA 63 D1
Arcadia WI 24 C1
Arcata CA 31 D1
Archbald PA 28 C4
Archbold OH 26 B4
Archdale NC 54 C2
Archer City TX 49 F4
Arco ID 19 F2
Arcola IL 40 A2
Arden CA 119 D2
Arden Hills MN 110 C1
Ardmore OK 50 B4
Ardmore PA 116 B3
Argentine MI 26 B3
Argo AL 53 D4
Argyle TX 62 A1
Arizona City AZ 56 A1
Arkadelphia AR 51 D4

Arkansas City KS 50 B1
Arkoma OK 51 D2
Arlee MT 10 A2
Arlington MA 96 D1
Arlington MN 24 A1
Arlington NE 23 F4
Arlington SD 23 E1
Arlington TN 52 B2
Arlington TX 62 A1
Arlington VT 29 D3
Arlington VA 68 A4
Arlington WA 8 B1
Arlington Hts. IL 98 B2
Arma KS 38 B4
Armona CA 44 B1
Armour SD 23 D2
Arnold CA 32 A3
Arnold MD 68 B4
Arnold MN 15 D3
Arroyo Grande CA 44 A3
Arroyo Hondo NM 48 A1
Arroyo Seco NM 48 A1
Artesia CA 106 D4
Artesia NM 58 A1
Arthur ND 13 F2
Arvada CO 35 F2
Arvin CA 44 C2
Asbury IL 25 D3
Asbury Park NJ 69 E2
Ashburn GA 65 F2
Ashburn VA 68 A4
Ashdown AR 51 D4
Asheboro NC 54 C2
Asheville NC 54 A2
Ashford AL 65 E3
Ash Grove MO 38 C4
Ashland AL 53 D4
Ashland CA 122 D3
Ashland KY 41 E3
Ashland MO 39 D3
Ashland MT 11 F4

Ashland NE 38 A1
Ashland OH 41 E1
Ashland OR 17 E3
Ashland PA 68 B1
Ashland VA 42 B4
Ashland WI 15 E3
Ashland City TN 52 C1
Ashley ND 13 E4
Ashtabula OH 27 E4
Ashton ID 20 A2
Ashville OH 41 D2
Aspen CO 35 E2
Aspen Hill MD 68 A4
Astoria OR 8 A3
Atascadero CA 44 A2
Atchison KS 38 B2
Athens AL 53 D3
Athens GA 53 F4
Athens OH 41 E2
Athens PA 28 B4
Athens TN 53 E2
Athens TX 62 B2
Athens WV 41 F4
Athol ID 9 F1
Athol MA 70 C1
Atkins AR 51 E2
Atkinson NE 23 D3
Atkinson NH 71 E1
Atlanta GA 53 F4
Atlanta TX 62 C1
Atlantic IA 24 A4
Atlantic Beach FL 66 B3
Atlantic City NJ 69 E3
Atlantic Highlands NJ 69 E2
Atmore AL 64 C3
Atoka OK 50 B3
Atoka TN 52 A2
Attalla AL 53 D4
Attica IN 40 A1
Attica NY 27 F3
Attleboro MA 71 E2
Atwater CA 32 A4

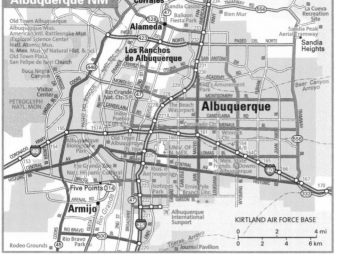

Albuquerque NM

Aberdeen MD 68 B3
Aberdeen MS 52 B4
Aberdeen NC 54 C3
Aberdeen SD 13 F4
Aberdeen WA 8 A3
Abernathy TX 49 D4
Abilene KS 37 F3
Abilene TX 59 E1
Abingdon IL 39 E1
Abingdon VA 54 B1
Abington MA 71 E2
Abita Sprs. LA 64 A3
Absarokee MT 11 D4
Absecon NJ 69 E3
Ackerman MS 52 B4
Ackley IA 24 B3
Acushnet MA 71 E3
Acworth GA 53 E4
Ada MN 14 A2
Ada OH 41 D1
Ada OK 50 B3
Adairsville GA 53 E3
Adams MA 70 B1
Adams WI 25 D2
Adamsville AL 53 D4
Adamsville TN 52 B2
Addis LA 63 F3
Addison IL 98 B4
Addison TX 100 F1
Adel GA 65 F3
Adel IA 24 A4
Adelanto CA 45 D3
Adelphi MD 124 D1
Adrian MI 26 B4
Adrian MO 38 B3
Affton MO 120 B3
Afton OK 50 C1
Afton WY 20 B3
Agawam MA 70 C2
Agua Fria NM 47 F2
Ahoskie NC 55 F1
Ahsahka ID 9 F3
Ahuimanu HI 72 A3
Aiken SC 54 B4
Ainsworth NE 22 C3
Airway Hts. WA 9 F1
Aitkin MN 14 C3
Ajo AZ 57 E4
Akiachak AK 74 A3
Akron CO 36 B2
Akron IA 23 F3
Akron NY 27 F3
Akron OH 41 F1
Akron PA 68 B2
Akutan AK 74 A4
Alabaster AL 53 D4
Alakanuk AK 74 A3
Alameda CA 122 B3
Alameda NM 47 F2
Alamo CA 122 D2
Alamo NM 47 E3
Alamo TN 52 B2
Alamogordo NM 57 F2
Alamosa CO 35 F4
Albany CA 122 B3
Albany GA 65 E2
Albany IN 40 C1
Albany KY 53 E1
Albany MN 14 B4

Albion IN 26 A4
Albion MI 26 B3
Albion NE 23 E4
Albion NY 27 F3
Albuquerque NM 47 F3
Alcester SD 23 F3
Alcoa TN 53 F2
Alcorn MS 63 F2
Alden NY 27 F3
Alderwood Manor WA 123 B2
Aldine TX 102 E1
Aledo IL 25 D4
Aledo TX 59 F1
Alexander City AL 65 D1
Alexandria IN 40 C1
Alexandria KY 41 D3
Alexandria LA 63 E2
Alexandria MN 14 B4
Alexandria SD 23 E2
Alexandria VA 68 A4
Alfred NY 28 A3
Algodones NM 47 F2
Algoma WI 25 F1
Algona IA 24 A3
Algonac MI 26 C3
Algood TN 53 E2
Alhambra CA 106 D2
Alice TX 60 C3
Aliceville AL 64 B1
Aliquippa PA 41 F1
Aliso Viejo CA 107 G5
Allegan MI 26 A3
Allen TX 62 A1
Allendale MI 26 A3
Allendale SC 66 B1
Allen Park MI 101 F3
Allentown PA 68 C1
Alliance NE 22 A4
Alliance OH 41 F1
Allyn WA 8 B2
Alma AR 51 D2
Alma GA 66 A2
Alma MI 26 B2
Alma NE 37 D2
Almont MI 26 C3
Alpena MI 16 C4
Alpharetta GA 53 F4
Alpine CA 56 B4
Alpine TX 58 B3
Alpine WY 20 B2
Alsip IL 98 D6
Alta IA 23 F3
Altadena CA 106 D1
Altamont KS 38 A4
Altamont OR 17 F3
Altamonte Sprs. FL 67 E1
Altavista VA 55 D1
Alton IL 39 E3
Altoona IA 24 B4
Altoona PA 42 B1
Altoona WI 24 C1
Altus OK 49 F3
Alum Creek WV 41 E3
Alva OK 49 F1
Alvarado TX 62 A1
Alvin TX 61 F1
Amarillo TX 49 D2
Ambler PA 69 D2
Amboy IL 25 E4

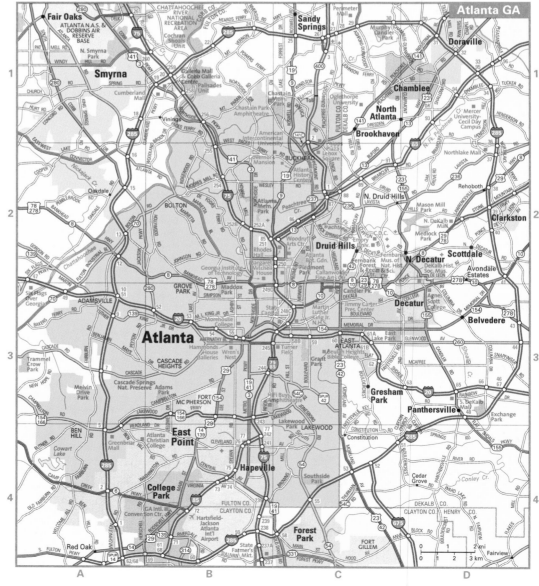

Atlanta GA

Entries in **bold color** indicate cities with detailed inset maps.

Atwood—Billerica **95**

Atwood KS	36 C2	Auburn KY	53 D1
Aubrey TX	50 B4	Auburn MA	71 D2
Auburn AL	65 D1	Auburn ME	29 F2
Auburn CA	31 F3	Auburn MI	26 B2
Auburn GA	53 F4	Auburn NE	38 A1
Auburn IL	39 F2	Auburn NY	28 B3
Auburn IN	26 B4	Auburn WA	8 B2
Auburn KS	38 A3	Auburndale FL	67 E2

Audubon IA	24 A4	Aurora CO	35 F2
Audubon NJ	116 D4	Aurora IL	25 E4
Augusta AR	51 F2	Aurora IN	40 C3
Augusta GA	54 A4	Aurora MN	15 D2
Augusta KS	37 F4	Aurora MO	51 D1
Augusta ME	29 F1	Aurora NE	37 E1
Ault CO	35 F1	Aurora OH	27 D4
Aumsville OR	17 E1	Aurora SD	23 E1

Austin IN	40 B3	Baldwinsville NY	28 B2
Austin MN	24 B2	Baldwyn MS	52 B3
Austin NV	33 D2	Ball LA	63 E2
Austin TX	59 F3	Ballinger TX	59 E2
Ava MO	51 E1	Ballston Spa NY	29 D3
Avalon CA	44 C4	Ballwin MO	39 E3
Avenal CA	44 B3	Balmville NY	70 A3
Aventura FL	109 B3	Baltic SD	23 E2
Avilla IN	26 B4	Baltimore MD	68 B4
Avoca IA	23 F4	Baltimore OH	41 E2
Avon CO	35 E2	Baltimore Highlands	
Avon NY	28 A3	MD	95 B3
Avon SD	23 E3	Bamberg SC	54 B4
Avondale AZ	46 B4	Bandon OR	17 D2
Avon Lake OH	27 D4	Bangor ME	30 B3
Avon Park FL	67 E2	Bangor PA	69 D1
Ayden NC	55 F3	Bangs TX	59 E2
Ayer MA	71 D1	Banning CA	45 D4
Azle TX	62 A1	Baraboo WI	25 D2
Aztec NM	47 F1	Barberton OH	41 E1
Azusa CA	106 E2	Barbourville KY	53 F1

B

Babbitt MN	15 D2	Bardstown KY	40 C4
Babylon NY	69 F1	Bargersville IN	40 B2
Bad Axe MI	26 C2	Bar Harbor ME	30 B4
Bagdad AZ	46 A3	Barling AR	51 D2
Baileys Crossroads		Barnesville GA	65 E1
VA	124 B3	Barnesville MN	14 A3
Bainbridge GA	65 D3	Barnesville OH	41 F2
Bainbridge Island		Barnhart MO	39 E3
WA	123 A3	Barnsdall OK	50 B1
Baird TX	59 E1	Barnstable MA	71 F3
Baker LA	63 F3	Bar Nunn WY	21 E3
Baker MT	12 B1	Barnwell SC	54 B4
Baker City OR	18 C1	Barracksville WV	42 A2
Bakersfield CA	44 B3	Barre VT	29 E1
Balch Sprs. TX	100 G3	Barrington RI	71 E3
Bald Knob AR	51 F2	Barron WI	15 D4
Baldwin GA	53 F4	Barrow AK	74 C1
Baldwin IL	39 E3	Barstow CA	45 D3
Baldwin LA	63 F4	Bartlesville OK	50 B1
Baldwin PA	117 G3	Bartlett IL	98 A3
Baldwin WI	24 C1	Bartlett TN	52 A2
Baldwin City KS	38 A3	Bartlett TX	62 A3
Baldwin Park CA	106 E2	Bartow FL	67 E2
		Barview OR	17 D2

Basalt CO	35 E2	Bellevue WA	8 B2
Basile LA	63 E3	Bellevue WI	25 F1
Basin WY	21 D1	Bellflower CA	106 D3
Bastrop LA	63 E1	Bell Gardens CA	106 D3
Bastrop TX	62 A4	Bellingham MA	71 D2
Batavia IL	25 E4	Bellingham WA	8 B1
Batavia NY	27 F3	Bellmawr NJ	116 D4
Batesburg-Leesville		Bellmead TX	62 A2
SC	54 B4	Bellows Falls VT	29 E2
Batesville AR	51 F2	Bells TN	52 B2
Batesville IN	40 C2	Bellville TX	62 B4
Batesville MS	52 A3	Bellwood IL	98 C4
Bath ME	29 F2	Belmar NJ	69 E2
Bath NY	28 A3	Belmond IA	24 B3
Bath PA	68 C1	Belmont CA	122 B5
Baton Rouge LA	63 F3	Belmont MA	96 D1
Battle Creek MI	26 B3	Belmont MS	52 C3
Battle Creek NE	23 E4	Belmont NC	54 B2
Battle Ground WA	8 B4	Beloit KS	37 E2
Battlement Mesa CO	35 D2	Beloit WI	25 E3
Battle Mtn. NV	33 D1	Belt MT	11 E1
Bawcomville LA	63 E1	Belton MO	38 B3
Baxley GA	66 A2	Belton SC	54 A3
Baxter MN	14 B3	Belton TX	62 A3
Baxter Sprs. KS	50 C1	Belvedere SC	54 A4
Bay AR	52 A2	Belvidere IL	25 E3
Bayard NE	22 A4	Belzoni MS	52 A4
Bayard NM	57 D1	Bemidji MN	14 B2
Bay City MI	26 B2	Benavides TX	60 C3
Bay City TX	61 E1	Benbrook TX	100 C3
Bayfield CO	35 D4	Bend OR	17 F2
Bay Minette AL	64 C3	Benicia CA	31 F4
Bayonet Pt. FL	67 D1	Benkelman NE	36 C2
Bayonne NJ	69 E1	Bennett CO	36 A2
Bayou Cane LA	63 F4	Bennettsville SC	54 C3
Bayou La Batre AL	64 B3	Bennington NE	23 F4
Bay Pt. CA	122 D1	Bennington VT	70 B1
Bay St. Louis MS	64 A3	Bensenville IL	98 C3
Bay Shore NY	69 F1	Benson AZ	56 B2
Bayshore Gardens FL	123 D5	Benson MN	14 A4
Bay Sprs. MS	64 A2	Benson NC	54 C2
Baytown TX	61 F1	Benton AR	51 E3
Bay Vil. OH	99 D2	Benton IL	39 F4
Beach ND	12 B3	Benton KY	52 C1
Beachwood NJ	69 E2	Benton LA	63 D1
Beachwood OH	99 G2	Benton City WA	9 D3
Beacon NY	70 A3	Benton Harbor MI	26 A3
Bear DE	68 C3	Bentonville AR	51 D1
Beardstown IL	39 E2	Berea KY	41 D4
Beatrice NE	37 F1	Berea OH	99 D3
Beatty NV	45 D1	Beresford SD	23 E3
Beaufort NC	55 F3	Bergenfield NJ	112 D1
Beaufort SC	66 B1	Berino NM	57 E2
Beaumont CA	45 D4	Berkeley CA	31 E4
Beaumont TX	62 C4	Berkeley MO	120 B1
Beaver OK	49 E1	Berkley MI	101 F1
Beaver PA	41 F1	Berlin CT	70 B3
Beaver UT	34 A3	Berlin MD	43 D3
Beaver Dam KY	40 B4	Berlin NH	29 E1
Beaver Dam WI	25 E2	Berlin NJ	69 D3
Beaver Falls PA	41 F1	Berlin WI	25 E2
Beaverton OR	8 B4	Bernalillo NM	47 F3
Becker MN	14 C4	Bernardsville NJ	69 D1
Beckley WV	41 F4	Berne IN	40 C1
Bedford IN	40 B3	Bernice LA	63 E1
Bedford IA	38 B1	Bernie MO	52 A1
Bedford MA	71 E1	Berryville AR	51 D1
Bedford OH	99 G3	Berryville VA	42 B3
Bedford PA	42 B2	Berthold ND	12 C1
Bedford TX	100 D2	Berthoud CO	35 F1
Bedford VA	44 A4	Berwick LA	63 F4
Bedford Hts. OH	99 G3	Berwick ME	29 F2
Beebe AR	51 F3	Berwick PA	42 C1
Beech Grove IN	103 C3	Berwyn IL	98 C4
Bee Ridge FL	67 D2	Bessemer AL	53 D4
Beeville TX	61 D2	Bessemer MI	15 E3
Beggs OK	50 B1	Bethalto IL	39 E3
Bel Air MD	68 B3	Bethany MO	38 B1
Belcourt ND	13 E1	Bethany OK	50 A2
Belding MI	26 B3	Bethel AK	74 B3
Belen NM	47 F3	Bethel CT	70 B3
Belfast ME	30 B4	Bethel OH	41 D3
Belfield ND	12 C3	Bethel VT	29 E2
Belgium WI	25 F2	Bethel Acres OK	50 B2
Belgrade MT	10 C4	Bethel Park PA	42 A1
Belington WV	42 A3	Bethesda MD	68 A4
Bell CA	106 D3	Bethlehem PA	68 C1
Bellair FL	103 D3	Bettendorf IA	25 D4
Bellaire OH	41 F2	Beulah CO	35 F4
Bellaire TX	61 E1	Beulah ND	12 C2
Bella Vista AR	51 D1	Beverly MA	71 E1
Belle WV	41 F3	Beverly Hills CA	106 C2
Belle Chasse LA	111 C2	Beverly Hills FL	67 D1
Bellefontaine OH	41 D1	Beverly Hills MI	101 F1
Bellefontaine Neighbors		Bexley OH	100 B2
MO	120 C1	Bicknell IN	40 A3
Bellefonte PA	42 B1	Biddeford ME	29 F2
Belle Fourche SD	21 F1	Big Bear City CA	45 D3
Belle Glade FL	67 F3	Big Bear Lake CA	45 D3
Belle Plaine IA	24 C4	Big Delta AK	74 C2
Belle Plaine KS	37 F4	Bigfork MT	10 A2
Belle Plaine MN	24 B1	Big Lake MN	14 C4
Belle Rose LA	63 F4	Big Lake TX	58 C3
Belleview FL	67 E1	Big Oak Flat CA	32 A4
Belleville IL	39 E3	Big Piney WY	20 B3
Belleville KS	37 E2	Big Rapids MI	26 A2
Belleville NJ	112 B2	Big Sandy MT	11 D1
Belleville WI	25 E2	Big Sky MT	10 C4
Bellevue ID	19 E2	Big Spr. TX	58 C1
Bellevue IA	25 D3	Big Stone City SD	14 A4
Bellevue NE	23 F4	Big Stone Gap VA	54 A1
Bellevue OH	26 C4	Big Timber MT	11 D4
Bellevue PA	117 F1	Billerica MA	71 E1

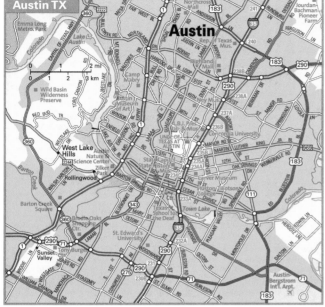

Austin TX

Baltimore MD

Figures after entries indicate page number and grid reference.

Entry	Pg	Grid
Billings MT	11	E4
Biloxi MS	64	B3
Binghamton NY	28	B4
Birch Bay WA	8	B1
Birdsboro PA	68	C2
Birmingham AL	53	D4
Birmingham MI	101	F1
Bisbee AZ	56	C2
Bishop CA	32	C4
Bishop TX	61	D3
Bishopville SC	54	C3
Bismarck MO	39	E4
Bismarck ND	13	D3
Bithlo FL	67	E1
Bixby OK	50	C2
Black Canyon City AZ	46	B4
Black Diamond WA	8	B2
Black Eagle MT	10	C2
Blackfoot ID	20	A2
Black Forest CO	36	A3
Black Hawk SD	22	A1
Blacklick Estates OH	100	B3
Black Mtn. NC	54	A2
Black River Falls WI	25	D1
Black Rock NM	47	D3
Blacksburg SC	54	B3
Blacksburg VA	41	F4
Blackshear GA	66	A2
Blackstone MA	71	D2
Blackstone VA	55	D1
Blackville SC	54	B4
Blackwell OK	50	B1
Blackwood NJ	69	D3
Blaine MN	14	C4
Blaine TN	53	F2
Blaine WA	8	B1
Blair NE	23	F4
Blairsville PA	42	A1
Blakely GA	65	E2
Blakely PA	28	C4
Blanchard LA	63	D1
Blanchard OK	50	A3
Blanchester OH	41	D2
Blanco TX	60	C2
Blanding UT	34	C4
Blissfield MI	26	B4
Bloomer WI	24	C1
Bloomfield CT	70	C3
Bloomfield IN	40	B3
Bloomfield IA	39	D1
Bloomfield MO	52	A1
Bloomfield NE	23	E3
Bloomfield NJ	112	A2
Bloomfield NM	47	F1
Bloomingdale IL	98	B3
Bloomingdale TN	54	A1
Blooming Prairie MN	24	B2
Bloomington CA	107	J2
Bloomington IL	39	F1
Bloomington IN	40	B3
Bloomington MN	24	B3
Bloomington TX	61	D2
Bloomsburg PA	42	C1
Blossom TX	50	C4
Blountsville AL	53	D4
Blountville TN	54	A1
Blue Ash OH	99	C1
Blue Earth MN	24	A2
Bluefield VA	41	F4
Bluefield WV	41	F4
Blue Hill NE	37	E1
Blue Island IL	98	D6
Blue Rapids KS	37	F2
Blue Ridge VA	42	A4
Blue Sprs. MO	38	B3
Bluff City TN	54	A1
Bluffton IN	40	C1
Bluffton OH	41	D1
Blythe CA	45	F4
Blytheville AR	52	A2
Boardman OH	41	F1
Boardman OR	9	D4
Boaz AL	53	D4
Boca Raton FL	67	F3
Boerne TX	59	F4
Bogalusa LA	64	A3
Boiling Spr. Lakes NC	55	D4
Boiling Sprs. NC	54	B2
Boiling Sprs. PA	68	A2
Boiling Sprs. SC	54	A3
Boise ID	19	D3
Boise City OK	48	C1
Boley OK	50	B2
Bolingbrook IL	25	F4
Bolivar MO	38	C4
Bolivar TN	52	B2
Bon Air VA	42	C4
Bondurant IA	24	B4
Bonham TX	50	B4
Bonifay FL	65	D3
Bonita Sprs. FL	67	E3
Bonner MT	10	B3
Bonners Ferry ID	9	F1
Bonner Sprs. KS	38	B3
Bonne Terre MO	39	E4
Bonney Lake WA	123	B5
Bono AR	51	F2
Bonsall CA	45	D4
Boone IA	24	B4
Boone NC	54	B1
Booneville AR	51	D2
Booneville MS	52	B3
Boonsboro MD	42	B2
Boonton NJ	69	E1
Boonville IN	40	A4
Boonville MO	38	C3
Boothbay Harbor ME	30	A4
Boothville LA	64	A4
Bordentown NJ	69	D2
Borger TX	49	D2
Boscobel WI	25	D3
Bosque Farms NM	47	F3
Bossier City LA	63	D1
Boston MA	71	E2
Bothell WA	8	B2
Bottineau ND	13	D1
Boulder CO	35	F2
Boulder MT	10	C3
Boulder City NV	45	F2
Boulder Creek CA	31	F4
Bound Brook NJ	69	E1
Bountiful UT	34	A1
Bourbonnais IL	25	F4
Bowbells ND	12	C1
Bowdle SD	13	E4
Bowie MD	68	B4
Bowie TX	50	A4
Bowling Green KY	53	D1
Bowling Green MO	39	D3
Bowling Green OH	26	C4
Bowman ND	12	B3
Box Elder MT	11	D1
Box Elder SD	22	A1
Boyertown PA	68	C2
Boyne City MI	16	B4
Boynton Beach FL	67	F3
Bozeman MT	10	C4
Brackettville TX	59	D4
Braddock Hts. MD	68	A3
Bradenton FL	67	D2
Bradford PA	27	F4
Bradford VT	29	E2
Bradley IL	25	F4
Bradley WV	41	F4
Brady TX	59	E3
Braidwood IL	25	E4
Brainerd MN	14	B3
Braintree MA	71	E2
Brandenburg KY	40	B4
Brandon FL	67	D2
Brandon MS	64	A1
Brandon SD	23	F2
Brandon VT	29	D2
Branford CT	70	C4
Branson MO	51	E1
Brant Rock MA	71	E2
Brattleboro VT	70	C1
Brawley CA	56	C3
Brazil IN	40	B3
Brazoria TX	61	E1
Brea CA	107	F3
Breaux Bridge LA	63	D2
Breckenridge CO	35	F2
Breckenridge MN	14	A3
Breckenridge TX	59	E2
Brecksville OH	99	F3
Breese IL	39	F3
Bremen GA	53	E4
Bremen IN	26	A4
Bremerton WA	8	B2
Brenham TX	62	B4
Brent AL	64	C1
Brentwood CA	31	F4
Brentwood NY	69	F1
Brentwood PA	117	G3
Brentwood TN	53	D2
Brevard NC	54	A2
Brewer ME	30	B3
Brewerton NY	28	B2
Brewster WA	9	D2
Brewton AL	64	C3
Briar TX	62	A1
Bridge City TX	63	D4
Bridgeport AL	53	E3
Bridgeport CT	70	B4
Bridgeport MI	26	B2
Bridgeport NE	22	A4
Bridgeport TX	50	A4
Bridgeport WA	9	D2
Bridgeport WV	41	F2
Bridger MT	11	E4
Bridgeton MO	120	A1
Bridgeton NJ	69	D3
Bridgetown OH	99	A2
Bridgeview IL	98	D5
Bridgeville DE	43	D3
Bridgewater MA	71	E2
Bridgewater SD	23	E2
Bridgewater VA	42	B3
Bridgman MI	26	A4
Bridgton ME	29	F2
Brigantine NJ	69	E3
Brigham City UT	20	A4
Brighton CO	35	F2
Brighton MI	26	C3
Brighton TN	52	A2
Brillion WI	25	E1
Brinkley AR	51	F3
Bristol CT	70	C3
Bristol NH	29	E2
Bristol RI	71	E3
Bristol TN	54	A1
Bristol VT	29	D2
Bristol VA	54	A1
Bristow OK	50	B2
Britt IA	24	B3
Britton SD	13	F4
Broad Brook CT	70	C2
Broadus MT	12	A4
Broadview Hts. OH	99	F3
Broadway VA	42	B3
Brockport NY	28	A2
Brockton MA	71	E2
Brodhead WI	25	E3
Broken Arrow OK	50	C2
Broken Bow NE	22	B4
Broken Bow OK	50	C4
Bronson MI	26	B4
Brookfield CT	70	B3
Brookfield IL	98	C4
Brookfield MO	38	C2
Brookfield WI	109	C2
Brookhaven MS	63	F2
Brookhaven NY	70	C4
Brookings OR	17	D4
Brookings SD	23	E1
Brookland AR	52	A2
Brookline MA	71	E2
Brooklyn OH	99	E2
Brooklyn Ctr. MN	110	B1
Brooklyn Park MD	95	B4
Brooklyn Park MN	110	B1
Brook Park OH	99	E3
Brooks KY	40	C4
Brookshire TX	61	E1
Brooksville FL	67	D1
Brooksville MS	52	B4
Brookville IN	40	C2
Brookville OH	40	C2
Brookville PA	27	F4
Broomall PA	116	A3
Broomfield CO	35	F2
Broussard LA	63	E4
Brown Deer WI	109	D1
Brownfield TX	58	C1
Browning MT	10	B1
Brownsburg IN	40	B2
Browns Mills NJ	69	D2
Brownstown IN	40	B3
Brownsville OR	17	E1
Brownsville, PA	42	A2
Brownsville TN	52	B2
Brownsville TX	61	D4
Brownwood TX	59	E2
Bruce MS	52	B4
Bruceton TN	52	C2
Bruceville-Eddy TX	62	A2
Brundidge AL	65	D2
Brunswick GA	66	B3
Brunswick ME	29	F3
Brunswick MD	68	A3
Brunswick OH	27	D4
Brush CO	36	A1
Brusly LA	63	F3
Bryan OH	26	B4
Bryan TX	62	B3

Birmingham AL (inset map)

Boston MA (inset map)

Entries in **bold color** indicate cities with detailed inset maps.

Bryant—Cathedral City **97**

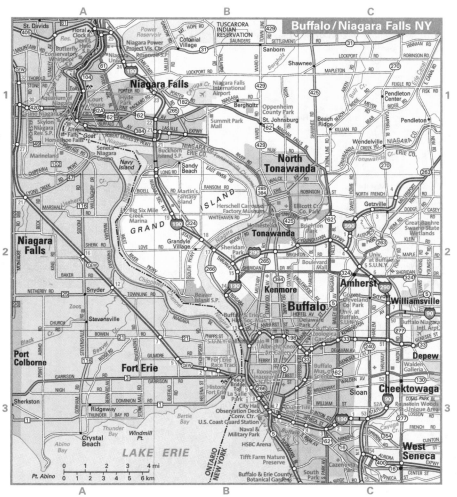

Buffalo/Niagara Falls NY

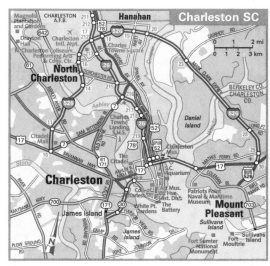

Charleston SC

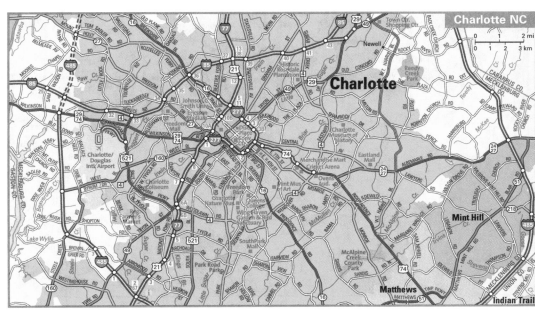

Charlotte NC

Bryant AR 51 E3	Burlington NC 54 C2	Calexico CA 56 C4
Bryn Mawr PA 69 D2	Burlington ND 13 D1	Calhoun GA 53 E3
Buchanan MI 26 A4	Burlington VT 29 D1	Calhoun City MS ... 52 B2
Buchanan Dam TX 59 F3	Burlington WA 8 B1	Calhoun Falls SC .. 54 A4
Buckeye AZ 46 A4	Burlington WI 25 E3	Caliente NV 33 E4
Buckhannon WV 41 F3	Burnet TX 59 F3	California MO 39 D3
Buckley WA 8 B2	Burnettown SC 54 B4	California PA 42 A2
Bucksport ME 30 B3	Burney CA 31 F1	California City CA 44 C3
Bucyrus OH 41 D1	Burns OR 18 B2	Calipatria CA 56 C3
Buda TX 59 F4	Burns Flat OK 49 F2	Calistoga CA 31 E3
Budd Lake NJ 69 D1	Burnside MN 110 B4	Callaway FL 65 D4
Buellton CA 44 B3	Burton MI 26 C3	Calumet City IL ... 98 E6
Buena NJ 69 D3	Burton SC 66 B1	Calvert TX 62 A3
Buena Park CA 106 C4	Burwell NE 23 D4	Calvert City KY ... 52 C1
Buena Vista CO 35 E3	Bushnell IL 39 E1	Calverton NY 70 C1
Buena Vista VA 42 A4	Bushyhead OK 50 C1	Camanche IA 25 D4
Buffalo MN 14 C4	Butler AL 64 A1	Camarillo CA 44 B3
Buffalo MO 38 C4	Butler IN 26 B4	Camas WA 8 B4
Buffalo NY 27 F3	Butler MO 38 B3	Cambria CA 44 A2
Buffalo OK 49 E1	Butler NJ 70 A4	Cambridge MD 43 D3
Buffalo TX 62 B2	Butler PA 42 A1	Cambridge MA 71 E2
Buffalo WV 41 E3	Butner NC 55 D2	Cambridge MN 14 C4
Buffalo WY 20 B1	Butte MT 10 B4	Cambridge NE 37 D1
Buffalo Grove IL ... 98 C2	Buzzards Bay MA .. 71 E3	Cambridge OH 41 E2
Buford GA 53 F4	Byers CO 36 A2	Cambridge City IN 40 C2
Buhl ID 19 E3	Bylas AZ 56 B1	Camden AL 64 C2
Buhler KS 37 F4	Byram MS 63 F2	Camden AR 51 E4
Buies Creek NC 55 D2	Byron GA 65 E1	Camden DE 68 C4
Bullhead City AZ .. 45 F3	Byron IL 25 E3	Camden ME 30 B4
Bull Shoals AR 51 E1	Byron MN 24 B2	Camden NJ 69 D2
Bulverde TX 59 F4	Byron WY 20 C1	Camden SC 54 C3
Buna TX 63 D3		Camden TN 52 C2
Bunker Hill OR 18 A2	**C**	Canadian TX 49 E2
Bunkerville NV 45 F1	Cabool MO 51 E1	
Bunkie LA 63 E3	Cabot AR 51 F3	
Buras LA 64 A4	Cache OK 49 F2	
Burbank CA 44 C4	Cactus TX 49 D2	
Burbank IL 98 D5	Cadillac MI 26 A1	
Burgaw NC 55 E3	Cadiz KY 52 C1	
Burien WA 8 B2	Cadiz OH 41 F1	
Burkburnett TX ... 49 F4	Cahokia IL 120 C3	
Burke SD 23 D2	Cairo GA 65 E3	
Burkesville KY 53 E1	Cairo IL 52 B1	
Burleson TX 62 A1	Calabasas CA 106 A2	
Burley ID 19 F3	Calais ME 30 C3	
Burlingame CA 122 B4	Calcium NY 28 B2	
Burlingame KS 38 A3	Caldwell ID 19 D2	
Burlington IA 39 E1	Caldwell KS 50 B1	
Burlington KS 38 A3	Caldwell TX 62 A3	
Burlington MA 71 E1	Caledonia MN 24 C2	
Burlington NJ 69 D2	Calera AL 64 C1	
Burlington KY 40 C3	Calera OK 50 B4	

Camarillo CA 44 B3	Camdenton MO 38 C4	
Camas WA 8 B4	Cameron LA 63 D4	
Cambria CA 44 A2	Cameron MO 38 B2	
Cambridge MD 43 D3	Cameron TX 62 A3	
Cambridge MA 71 E2	Cameron WV 41 F2	
Cambridge MN 14 C4	Cameron WI 15 D4	
Cambridge NE 37 D1	Camilla GA 65 E2	
Cambridge OH 41 E2	Campbell CA 31 F4	
Cambridge City IN 40 C2	Campbell MO 52 A1	
Camden AL 64 C2	Campbellsport WI .. 25 E2	
Camden AR 51 E4	Campbellsville KY . 40 C4	
Camden DE 68 C4	Camp Hill PA 68 A2	
Camden ME 30 B4	Campion CO 35 F1	
Camden NJ 69 D2	Camp Sprs. MD ... 124 D3	
Camden SC 54 C3	Camp Verde AZ 46 B3	
Camden TN 52 C2	Canadian TX 49 E2	

Canal Fulton OH 41 E1	Carefree AZ 46 B4	Carrollton TX 62 A1
Canal Winchester OH . 41 E2	Carencro LA 63 E3	Carroll Valley PA .. 68 A3
Canandaigua NY 28 A3	Carey ID 19 E2	Carrollwood FL 123 E1
Canastota NY 28 B3	Carey OH 41 D1	Carson CA 106 C4
Canby MN 23 F1	Caribou ME 30 C1	Carson City NV 32 B2
Canby OR 8 B4	Carleton MI 26 C4	Carteret NJ 112 A5
Cando ND 13 E1	Carlin NV 33 D1	Cartersville GA 53 E4
Caney KS 50 C1	Carlinville IL 39 E2	Carterville IL 39 F4
Canfield OH 41 F1	Carlisle AR 51 F3	Carthage IL 39 D1
Canistota SD 23 E2	Carlisle IN 40 A3	Carthage MS 64 A1
Cannon Ball ND 13 D2	Carlisle IA 24 A3	Carthage MO 38 B4
Cannon Beach OR ... 8 A4	Carlisle KY 41 D3	Carthage NY 28 C2
Cannon Falls MN 24 B1	Carlisle PA 68 A2	Carthage TN 53 D1
Canon City CO 35 F3	Carlsbad CA 56 A3	Carthage TX 62 C2
Canonsburg PA 41 F1	Carlsbad NM 58 A1	Caruthersville MO . 52 B2
Canton GA 53 E4	Carlton OR 8 A4	Cary IL 25 E3
Canton IL 39 E1	Carlton MN 14 C3	Cary NC 55 D2
Canton MS 64 A1	Carlyle IL 39 F3	Caryville TN 53 F1
Canton MO 39 D2	Carlyss LA 63 D4	Casa Blanca NM ... 47 E3
Canton NY 28 C1	Carmel IN 40 B2	Casa Grande AZ ... 56 A1
Canton NC 54 A2	Carmel NY 70 B3	Cascade CO 35 F3
Canton OH 41 F1	Carmel-by-the-Sea CA 31 D4	Cascade ID 19 D1
Canton SD 23 F2	Carmel Valley CA .. 31 D4	Cascade IA 24 C3
Canton TX 62 B1	Carmi IL 40 A4	Cascade MT 10 C2
Canutillo TX 57 E2	Carnation WA 8 B2	Casey IL 40 A2
Canyon TX 49 D3	Carnegie OK 49 F3	Casey IL 40 A2
Canyon Day AZ 46 C4	Carney MD 95 C1	Casper WY 21 E3
Cape Canaveral FL .. 67 F1	Caro MI 26 C2	Cass City MI 26 C2
Cape Coral FL 67 E3	Carol City FL 67 F4	Casselberry FL 115 C1
Cape Elizabeth ME .. 29 F2	Carolina Beach NC . 55 E4	Casselton ND 13 F3
Cape Girardeau MO . 39 F4	Carol Stream IL ... 98 A4	Cassville MO 51 D1
Cape May NJ 69 D4	Carpentersville IL . 25 E3	Castaic CA 44 C3
Cape May C.H. NJ .. 69 D4	Carpinteria CA 44 B3	Castle Dale UT 34 B2
Cape St. Claire MD . 68 B4	Carrington ND 13 E2	Castle Rock CO 35 F2
Capitan NM 48 A1	Carrizo Sprs. TX .. 60 B2	Castle Rock WA ... 8 B3
Capitola CA 31 D4	Carrizozo NM 48 A4	Castle Shannon PA 117 F3
Captain Cook HI 73 E3	Carroll IA 24 A4	Castlewood SD 23 E1
Caraway AR 52 A2	Carrollton GA 53 E4	Castlewood VA 54 A1
Carbondale CO 35 E2	Carrollton IL 39 E2	Castro Valley CA .. 122 D3
Carbondale IL 39 F4	Carrollton KY 40 C3	Castroville CA 31 D4
Carbondale KS 38 A3	Carrollton MI 26 B2	Castroville TX 59 F4
Carbondale PA 28 C4	Carrollton MO 38 C2	Catalina AZ 56 B1
Carbon Hill AL 52 C4	Carrollton OH 41 F1	Catasauqua PA 68 C1
		Cathedral City CA 45 D4

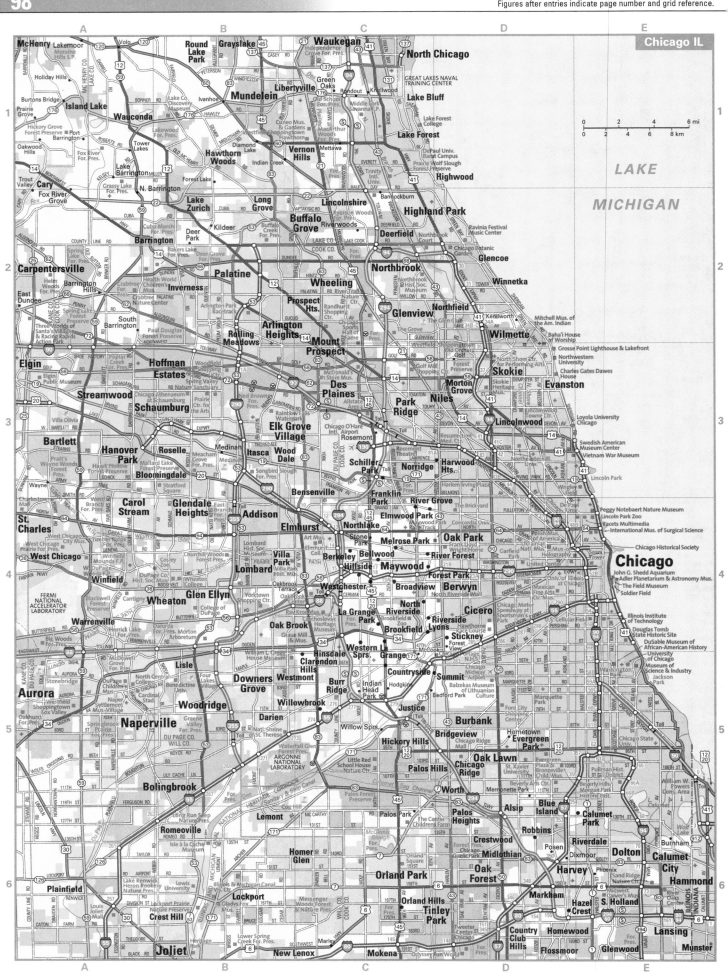

Chicago IL

LAKE MICHIGAN

Entries in **bold color** indicate cities with detailed inset maps.

Catlettsburg KY 41 E3
Catonsville MD 68 B4
Catoosa OK 50 C2
Catskill NY 70 A2
Cavalier ND 13 F1
Cave City AR 51 F2
Cave City KY 53 D1
Cave Creek AZ 46 B4
Cave Jct. OR 17 D3
Cayce SC 54 B4
Cayuga Hts. NY 28 B3
Cazenovia NY 28 B3
Cecilia LA 63 E3

Cedarburg WI 25 F2
Cedar City UT 33 F4
Cedaredge CO 35 D3
Cedar Falls IA 24 C3
Cedar Grove FL 65 D4
Cedar Grove NJ 112 A4
Cedar Grove NM 47 F2
Cedar Grove WI 25 F2
Cedar Hill MO 39 E3
Cedar Hill TX 100 E3
Cedar Hills OR 118 A4
Cedar Lake IN 25 F4
Cedar Park TX 59 F3

Cedar Rapids IA 24 C4
Cedar Sprs. MI 26 A2
Cedartown GA 53 E4
Cedarville OH 41 D2
Celina OH 40 C1
Celina TX 50 B4
Centennial CO 101 D3
Center CO 35 F4
Center ND 13 D3
Center TX 62 C2
Center Pt. AL 53 D4

Center Pt. IA 24 C3
Centerton AR 51 D1
Centerville AR 51 E2
Centerville GA 65 F1
Centerville IN 40 C2
Centerville IA 38 C1
Centerville MA 71 F3
Centerville SD 23 E3
Centerville TN 52 C2
Centerville UT 34 A1
Centerfield UT 34 A3
Center Pt. AL 53 D4

Central Falls RI 71 D2
Centralia IL 39 F3
Centralia MO 39 D2
Centralia WA 8 B3
Central Park WA 8 A3
Central Pt. OR 17 E3
Centre AL 53 E4
Centreville AL 64 C1
Centreville MS 63 F3
Centreville VA 68 A4
Ceres CA 31 F4
Ceresco NE 37 F1
Cerritos CA 106 E4

Chackbay LA 63 F4
Chadbourn NC 55 D3
Chadron NE 22 A3
Chaffee MO 39 F4
Chagrin Falls OH 27 D4
Chalfant PA 69 D2
Chalmette LA 64 A4
Chama NM 47 F1
Chamberlain SD 23 D2
Chambersburg PA 42 B2
Chamblee GA 94 C1
Champaign IL 40 A2

Chandler AZ 46 B4
Chandler IN 40 A4
Chandler OK 50 B2
Chandler TX 62 B1
Chanhassen MN 24 B1
Channahon IL 25 E4
Channelview TX 102 F3
Chantilly VA 68 A4
Chanute KS 38 A4
Chaparral NM 57 E3
Chapel Hill NC 55 D2
Chapman KS 37 F3
Chapmanville WV 41 E4

Chappell NE 36 B1
Charenton LA 63 F4
Chariton IA 38 C1
Charles City IA 24 B3
Charleston AR 51 D2
Charleston IL 40 A3
Charleston MS 52 A4
Charleston MO 52 B1
Charleston SC 66 C1
Charleston WV 41 F3
Charles Town WV 42 B2
Charlevoix MI 16 B3
Charlo MT 10 A2

Chandler AZ 46 B4
Chisholm MN 14 C2
Chittenango NY 28 B3
Choctaw OK 50 B2
Choteau MT 10 C2
Chouteau OK 50 C2
Chowchilla CA 31 F3
Christiansburg VA 41 F4
Christopher IL 39 F4
Chubbuck ID 19 F3
Chula Vista CA 56 B4
Church Hill TN 54 A3
Church Pt. LA 63 E3
Church Rock NM 47 D2

Charlotte MI 26 B3
Cicero IL 25 F4
Charlotte NC 54 B2
Cicero IN 40 B2
Charlotte TX 60 C1
Cimarron KS 37 D4
Charlottesville VA 42 B4
Cimarron NM 48 A1
Chase City VA 55 D1
Cincinnati OH 40 C3
Chaska MN 24 B1
Circle MT 12 A2
Chatfield MN 24 C2
Circleville OH 41 E2
Chatham IL 39 F2
Cisco TX 59 E1
Chatsworth GA 53 E3
Citronelle AL 64 B3
Chattahoochee FL 65 E3
Citrus Hts. CA 31 F3
Chattanooga TN 53 E3
Citrus Park FL 123 D1
Chauvin LA 63 F4
Citrus Spgs. FL 67 D1
Cheboygan MI 16 C4
Clairton PA 42 A1
Checotah OK 50 C2
Clancy MT 10 C3
Cheektowaga NY 27 F3
Clanton AL 65 D1
Chehalis WA 8 B3
Clare MI 26 B2
Chelan WA 8 C2
Claremont CA 107 G2
Chelmsford MA 71 D1
Claremont NH 29 E2
Chelsea AL 53 D4
Claremore OK 50 C1
Chelsea MA 96 F1
Clarence IA 24 B3
Chelsea MI 26 B3
Clarion IA 24 B3
Chelsea OK 50 C1
Clarion PA 42 A1
Cheney KS 37 F4
Clark SD 23 E1
Cheney WA 9 E2
Clarkdale AZ 46 B3
Chenoweth OR 8 C4
Clark Fork ID 9 F1
Cheraw SC 54 C3
Clarksburg WV 41 F2
Cherokee IA 23 F3
Clarksdale MS 52 A3
Cherokee OK 49 F1
Clarks Summit PA 28 B4
Cherokee Vil. AR 51 F1
Clarkston WA 9 F3
Cherry Hill NJ 116 E4
Clarksville AR 51 D2
Cherryvale KS 38 A4
Clarksville IN 108 A1
Cherryvale SC 54 C4
Clarksville TN 52 C1
Cherryville NC 54 B2
Clarksville TX 50 C4
Chesaning MI 26 B2
Clatskanie OR 8 A3
Chesapeake VA 55 F1
Claxton GA 54 C4
Chesapeake WV 41 F3
Clay Ctr. KS 37 F2
Chesapeake Beach
Clay Ctr. NE 37 E1
   MD 42 C3
Claypool AZ 46 C4
Cheshire CT 70 C3
Clayton CA 53 F3
Chester IL 39 E4
Clayton DE 68 C4
Chester MD 68 B4
Clayton MO 39 E3
Chester MT 10 C1
Clayton NJ 69 D3
Chester NY 70 A3
Clayton NM 48 C1
Chester PA 68 C3
Clayton NC 55 D2
Chester SC 54 B3
Clearfield PA 42 B1
Chester VT 29 E2
Clearfield UT 20 A4
Chester VA 42 C4
Clearlake CA 31 E3
Chester WV 41 F1
Clear Lake IA 24 B2
Chesterfield MO 39 E3
Clear Lake SD 23 E1
Chesterfield VA 42 C4
Clearwater FL 67 D2
Chesterton IN 25 E4
Clearwater SC 37 F4
Chestertown MD 68 C4
Cleburne TX 62 A1
Chestnut Ridge NY 70 A4
Cle Elum WA 8 C2
Chetek WI 15 D4
Clemmons NC 54 C2
Chetopa KS 50 C1
Clemson SC 54 A3
Chevak AK 74 A3
Clendenin WV 41 F3
Cheviot OH 99 A2
Clermont FL 67 E1
Chewelah WA 9 F1
Cleveland FL 67 E3
Cheyenne WY 21 F4
Cleveland OH 27 D4
Cheyenne Wells CO 36 B3
Cleveland MS 52 A4
Chicago IL 25 F4
Cleveland OK 50 B1
Chickamauga GA 53 E3
Cleveland TN 53 E3
Chickasaw AL 64 B3
Cleveland TX 62 C3
Chickasha OK 50 A3
Cleveland Hts. OH 27 D4
Chico CA 31 F2
Clewiston FL 67 E3
Chicopee MA 70 C2
Cliffside Park NJ 112 D2
Childersburg AL 53 D4
Clifton AZ 56 C1
Childress TX 49 E3
Clifton CO 35 D3
Chilhowie VA 54 B1
Clifton NJ 69 E1
Chillicothe IL 39 F1
Clifton TN 52 C2
Chillicothe MO 38 C2
Clifton TX 62 A2
Chillicothe OH 41 E2
Clifton Forge VA 42 A4
Chillum MD 124 D1
Clifton Park NY 70 A1
Chilton WI 25 E2
Clinton AR 51 E2
Chimayo NM 47 F2
Clinton CT 70 C4
China Grove NC 54 C2
Clinton IL 39 F1
Chincoteague VA 43 D4
Clinton IN 40 A3
Chinle AZ 47 D2
Clinton IA 25 D4
Chino CA 107 G3
Clinton KY 52 B1
Chino Hills CA 107 G3
Clinton LA 63 F3
Chino Valley AZ 46 B3
Clinton ME 30 B3
Chinook MT 11 D1
Clinton MD 68 A4
Chipley FL 65 D3
Clinton MA 71 D2
Chippewa Falls WI 24 C1
Clinton MI 26 B4
Chisholm ME 29 F1
Clinton MS 63 F1
Clinton MO 38 C3
Clinton NC 55 D3
Clinton OK 49 F2
Clinton SC 54 B4
Clinton TN 53 F2
Clintonville WI 25 E1
Clintwood VA 54 A1

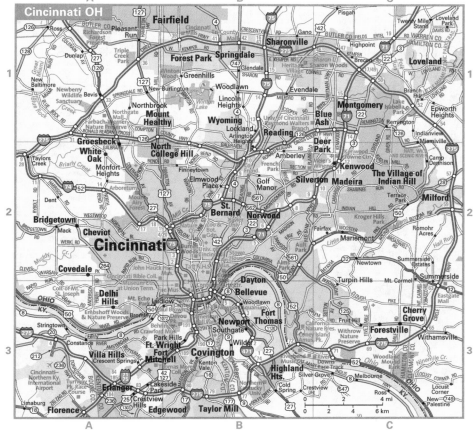

Cincinnati OH

Cleveland OH

Clio AL 65 D2
Clio MI 26 C2
Cloquet MN 15 D3
Cloudcroft NM 57 F1
Clover SC 54 B3
Cloverdale CA 31 E3
Cloverdale IN 40 B2
Cloverleaf TX 102 F2
Clovis CA 44 B1
Clovis NM 48 C3
Clute TX 61 F1

Clyde OH 26 C4
Clyde TX 59 E1
Coachella CA 45 E4
Coal City IL 25 E4
Coalgate OK 50 B3
Coalinga CA 44 A2
Coalville UT 34 B1
Coatesville PA 68 C2
Cobleskill NY 70 A1
Cochiti NM 47 C3
Cochran GA 65 F1

Cockeysville MD 68 B3
Cocoa FL 67 F1
Cocoa Beach FL 67 F1
Coconut Creek FL 109 B1
Cody WY 20 C1
Coeburn VA 54 A1
Coeur d'Alene ID 9 E1
Coffeyville KS 50 C1
Cohasset MA 71 E2
Cohasset MN 14 C2
Cohoes NY 70 B1

Cokato MN 24 A1
Cokeville WY 20 B4
Colby KS 36 C2
Colby WI 25 D1
Colchester CT 70 C2
Cold Spr. MN 14 B4
Coldwater MI 26 B4
Coldwater MS 52 A3
Coldwater OH 40 C1
Coleman TX 59 E2
Colfax IA 24 B3

Colfax LA 63 E2
Colfax WA 9 E3
Collegedale TN 53 E3
College AK 74 C2
College Park GA 94 B4
College Park MD 124 E1
College Place WA 9 D4
College Sta. TX 62 B3
Collierville TN 52 A3
Collingdale PA 116 B4
Collingswood NJ 116 D4
Collins MS 64 A2
Collinsville AL 53 D3
Collinsville IL 120 D2
Collinsville MS 64 B1
Collinsville OK 50 C1
Collinsville VA 54 C1
Colman SD 23 E2
Colonial Beach VA 42 C4
Colonial Hts. VA 42 C4
Colonie NY 70 A1
Colorado City AZ 46 A3
Colorado City CO 35 F4
Colorado City TX 59 D1
Colorado Sprs. CO 35 F3
Colstrip MT 11 F4
Colton CA 107 J2
Colton SD 23 E2
Columbia KY 53 E1
Columbia MD 68 A4
Columbia MS 64 A2
Columbia MO 39 D3
Columbia PA 68 C1
Columbia SC 54 B4
Columbia TN 53 D2
Columbia City IN 26 A4
Columbia City OR 8 B4
Columbia Falls MT 10 A1
Columbia Hts. MN 110 C2
Columbiana AL 65 D1
Columbiana OH 41 F1
Columbus GA 65 E1
Columbus IN 40 B2
Columbus KS 38 B4
Columbus MS 52 B4
Columbus MT 11 E4
Columbus NE 23 E4
Columbus NM 57 E4
Columbus OH 41 D2
Columbus TX 61 E1
Columbus WI 25 E2
Columbus Jct. IA 24 C4
Colusa CA 31 F2
Colville WA 9 E1
Colwich KS 37 F4
Comanche OK 50 A3
Comanche TX 59 F2
Combes TX 61 D4
Comfort TX 59 F4

Commack NY 69 F1
Commerce GA 53 F3
Commerce OK 50 C1
Commerce TX 50 C4
Commerce City CO 101 D2
Como MS 52 A3
Compton CA 44 C4
Concord CA 31 F4
Concord MA 71 D1
Concord MO 120 B3
Concord NH 29 E2
Concord NC 54 C2
Concordia KS 37 F2
Concordia MO 38 C3
Congress AZ 46 A4
Conneaut OH 27 E4
Connell WA 9 D3
Connellsville PA 42 A2
Connersville IN 40 C2
Conover NC 54 B2
Conrad MT 10 C1
Conroe TX 62 B3
Constantine MI 26 A4
Conway AR 51 E3
Conway NH 29 F2
Conway SC 55 D4
Conway Sprs. KS 37 F4
Conyers GA 53 F4
Cookeville TN 53 E2
Coolidge AZ 56 A1
Coon Rapids MN 14 C4
Cooper TX 50 C4
Coopersburg PA 68 C1
Cooperstown ND 13 F2
Coos Bay OR 17 D2
Coppell TX 100 E1
Copperas Cove TX 59 F3
Coquille OR 17 D2
Coral Gables FL 67 F4
Coral Hills MD 124 C2
Coral Sprs. FL 67 F3
Coralville IA 24 C4
Coram NY 69 F1
Corbin KY 53 F1
Corcoran CA 44 B2
Cordele GA 65 F2
Cordell OK 49 F2
Cordova AL 52 C4
Cordova AK 74 C3
Corinth MS 52 B3
Cornelia GA 53 F3
Cornelius NC 54 B2
Corning AR 52 A1
Corning CA 31 E2
Corning IA 38 B1
Corning NY 28 B3
Cornville AZ 46 B3
Cornwall PA 68 B2
Cornwall-on-Hudson
  NY 70 A3

Corona CA 45 D4
Coronado CA 56 B4
Corpus Christi TX 61 D3
Corrales NM 47 F2
Corrigan TX 62 C3
Corry PA 42 A1
Corsica SD 23 D2
Corsicana TX 62 B2
Cortaro AZ 56 B1
Cortez CO 35 D4
Cortland NY 28 B3
Cortland OH 27 E4
Corunna MI 26 B3
Corvallis OR 17 E1
Corvallis MT 10 B3
Coshocton OH 41 E1
Cosmopolis WA 8 A3
Costa Mesa CA 106 E5
Cotati CA 31 E3
Cottage Grove MN 110 E4
Cottage Grove OR 17 E2
Cottonport LA 63 E3
Cottonwood AZ 46 B3
Cottonwood ID 9 F4
Cottonwood Hts. UT ... 120 B2
Cotulla TX 60 B2
Coudersport PA 27 F4
Council ID 19 D1
Council Bluffs IA 23 F4
Council Grove KS 38 A3
Country Club Hills IL ... 98 D6
Country Homes WA 9 E2
Coupeville WA 8 B1
Courtney TX 62 B3
Coushatta LA 63 D2
Covina CA 107 F2
Covington GA 53 F4
Covington IN 40 A1
Covington KY 40 C3
Covington LA 64 A3
Covington OH 41 D2
Covington TN 52 B2
Covington VA 42 A4
Covington WA 123 B4
Cowan TN 53 D2
Coweta OK 50 C2
Cowley WY 20 C1
Cowpens SC 54 B3
Coxsackie NY 70 A2
Cozad NE 37 D1
Craig AK 75 E4
Craig CO 35 D1
Craigmont ID 9 F3
Craigsville WV 41 F3
Crandon WI 15 F3
Crane TX 58 C2
Cranston RI 71 D3
Crawford NE 22 A1
Crawfordsville IN 40 B2
Creedmoor NC 55 D2
Creighton NE 23 E3

Creola AL 64 B3
Cresaptown MD 42 B3
Crescent OK 50 A2
Crescent City CA 17 D4
Cresco IA 24 C2
Crested Butte CO 35 E3
Crestline CA 45 D3
Crestline OH 41 E1
Creston IA 38 B1
Crestview FL 65 D3
Crestwood KY 40 C3
Crestwood MO 120 B3
Creswell OR 17 E2
Crete IL 25 F4
Crete NE 37 F1
Creve Coeur MO 120 A2
Crewe VA 42 B4
Crimora VA 42 B3
Cripple Creek CO 35 F3
Crisfield MD 43 D4
Crittenden KY 40 C3
Crockett TX 62 B2
Crofton MD 68 B4
Crooks SD 23 E2
Crookston MN 14 A2
Crosby MN 14 C3
Crosby ND 12 B1
Crosby TX 62 C4
Crosbyton TX 49 D4
Crossett AR 63 E1
Crosslake MN 14 B3
Crossville TN 53 E2
Croswell MI 26 C2
Croton-on-Hudson NY ... 70 A4
Crow Agency MT 11 F4
Crowley LA 63 E3
Crowley TX 62 A1
Crown Hts. NY 70 A3
Crown Pt. IN 25 F4
Crownpoint NM 47 E2
Crump TN 52 C2
Crystal MN 110 B2
Crystal City MO 39 E3
Crystal City TX 60 B2
Crystal Lake IL 25 E3
Crystal River FL 67 D1
Crystal Sprs. MS 63 F2
Cuba MO 39 D4
Cuba NM 47 F2
Cuba City WI 25 D3
Cudahy CA 106 D3
Cudahy WI 25 F3
Cuero TX 61 D1
Culbertson MT 12 B1
Cullman AL 53 D4
Cullowhee NC 53 F2
Culpeper VA 42 B3
Culver City CA 106 B3
Cumberland KY 54 A1
Cumberland MD 42 B3
Cumberland WI 15 D4
Cumming GA 53 F3

Columbus OH

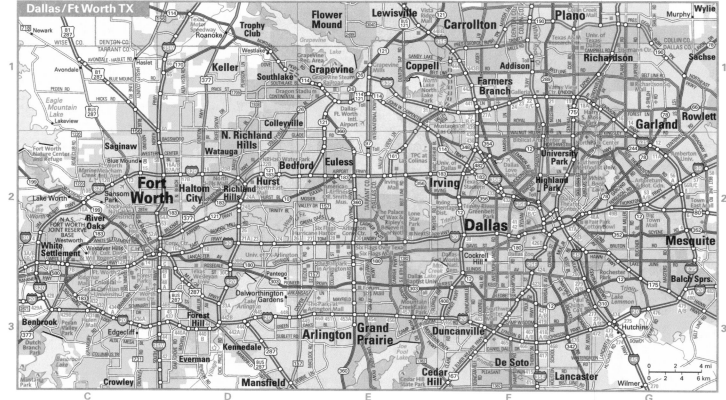

Dallas / Ft Worth TX

Entries in **bold color** indicate cities with detailed inset maps.

Cupertino—De Smet **101**

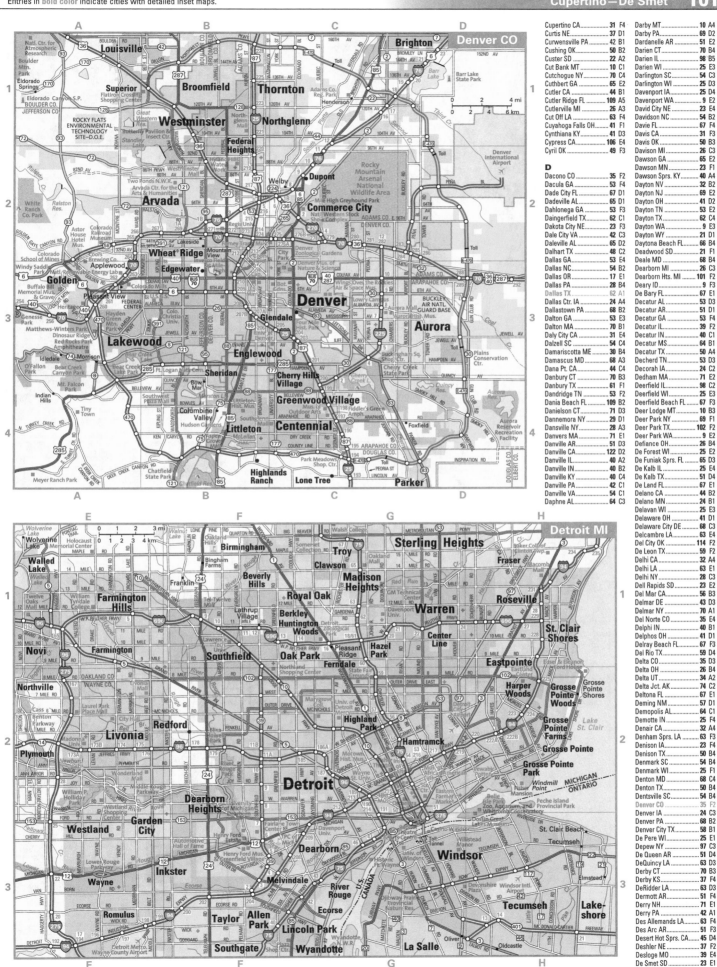

Cupertino CA	31	F4
Curtis NE	37	D1
Curwensville PA	42	B1
Cushing OK	50	B2
Custer SD	22	A2
Cut Bank MT	10	C1
Cutchogue NY	70	C4
Cuthbert GA	65	E2
Cutler CA	32	B2
Cutler Ridge FL	109	A5
Cutlerville MI	26	A3
Cut Off LA	63	F4
Cuyahoga Falls OH	41	F1
Cynthiana KY	41	D3
Cypress CA	106	E4
Cyril OK	49	F3
D		
Dacono CO	35	F2
Dacula GA	53	F3
Dade City FL	67	D1
Dadeville AL	65	D1
Dahlonega GA	53	F3
Daingerfield TX	62	C1
Dakota City NE	23	F4
Dale City VA	42	C3
Dalhart TX	48	C2
Dallas GA	53	E4
Dallas NC	54	B2
Dallas OR	17	E1
Dallas PA	28	B4
Dallas TX	62	A1
Dallas Ctr. IA	24	A4
Dallastown PA	68	B2
Dalton GA	53	E3
Dalton MA	70	B1
Dalzell SC	54	C4
Daly City CA	31	E4
Damariscotta ME	30	B4
Damascus MD	68	A3
Dana Pt. CA	44	C4
Danbury CT	70	B3
Danbury TX	61	F1
Dandridge TN	53	F2
Dania Beach FL	109	B2
Danielson CT	71	D3
Dannemora NY	29	D1
Dansville NY	28	A3
Danvers MA	71	E1
Danville AR	51	D3
Danville CA	122	B3
Danville IL	40	A2
Danville IN	40	B2
Danville KY	40	C4
Danville PA	42	C1
Danville VA	54	C1
Daphne AL	64	C3
Darby MT	10	A4
Darby PA	69	D2
Dardanelle AR	51	E2
Darien CT	70	B4
Darien IL	98	B5
Darien WI	25	E3
Darlington SC	54	C3
Darlington WI	25	D3
Darrington WA	25	D4
Davenport WA	9	E2
David City NE	23	E4
Davidson NC	54	B2
Davie FL	67	F4
Davis CA	31	F3
Davis OK	50	B3
Davison MI	26	C2
Dawson GA	65	E2
Dawson MN	23	F1
Dawson Sprs. KY	40	A4
Dayton NV	32	B2
Dayton NJ	69	E2
Dayton OH	41	D2
Dayton TN	53	E2
Dayton TX	62	C4
Dayton WA	9	E3
Dayton WY	21	D1
Daytona Beach FL	66	B4
Deadwood SD	21	F1
Deale MD	68	B4
Dearborn MI	26	C3
Dearborn Hts. MI	101	F2
Deary ID	9	F3
De Bary FL	67	E1
Decatur AL	53	D3
Decatur AR	51	D1
Decatur GA	53	F3
Decatur IL	39	F2
Decatur IN	40	C1
Decatur MS	64	B1
Decatur TX	50	A4
Decherd TN	53	D3
Decorah IA	24	C2
Dedham MA	71	E2
Deerfield IL	98	C2
Deerfield WI	25	E3
Deerfield Beach FL	67	F3
Deer Lodge MT	10	B3
Deer Park NY	69	F1
Deer Park TX	102	F2
Deer Park WA	9	E2
Defiance OH	26	B4
De Forest WI	25	E2
De Funiak Sprs. FL	65	D3
De Kalb IL	25	E4
De Kalb TX	51	D4
De Land FL	67	E1
Delano CA	44	B2
Delano MN	24	B1
Delavan WI	25	E3
Delaware OH	41	D1
Delaware City DE	68	C3
Delcambre LA	63	E4
Del City OK	114	F2
De Leon TX	59	F2
Delhi CA	32	A4
Delhi LA	63	E1
Delhi NY	28	C3
Dell Rapids SD	23	E2
Del Mar CA	56	B3
Delmar DE	43	D3
Delmar NY	70	A1
Del Norte CO	35	E4
Delphi IN	40	B1
Delphos OH	41	D1
Delray Beach FL	67	F3
Del Rio TX	59	D4
Delta CO	35	D3
Delta OH	26	B4
Delta UT	34	A2
Delta Jct. AK	74	C2
Deltona FL	67	E1
Deming NM	57	D1
Demopolis AL	64	C1
Demotte IN	25	F4
Denair CA	32	A4
Denham Sprs. LA	63	E3
Denison IA	23	F4
Denison TX	50	B4
Denmark SC	54	B4
Denmark WI	25	F1
Denton MD	68	C4
Denton TX	50	B4
Dentsville SC	54	B4
Denver CO	35	F2
Denver IA	24	C3
Denver PA	68	B2
Denver City TX	58	B1
De Pere WI	25	E1
Depew NY	97	C3
De Queen AR	51	D4
DeQuincy LA	63	D3
Derby CT	70	B3
Derby KS	37	F4
Desloge MO	39	E4
DeRidder LA	51	F4
Dermott AR	51	F4
Derry NH	71	E1
Derry PA	42	A1
Des Allemands LA	63	F4
Des Arc AR	51	F3
Desert Hot Sprs. CA	45	D4
Deshler NE	37	F2
Desloge MO	39	E4
De Smet SD	23	E1

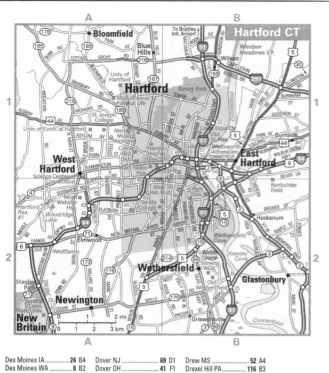

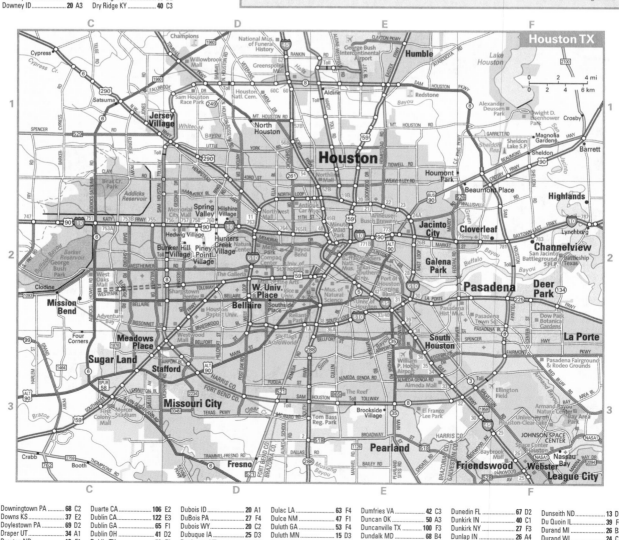

Des Moines IA	24	B4
Des Moines WA	8	B2
De Soto KS	38	B3
De Soto MO	39	E4
De Soto TX	62	A1
Des Peres MO	120	A2
Des Plaines IL	98	C3
Destin FL	65	D3
Destrehan LA	111	A2
Detroit MI	26	C3
Detroit Lakes MN	14	A3
Devils Lake ND	13	E2
Devine TX	60	C1
Dewey AZ	46	B3
Dewey OK	50	C1
De Witt AR	51	F3
De Witt IA	25	D4
De Witt MI	26	B3
Dexter ME	30	B3
Dexter MI	26	B3
Dexter MO	52	A1
Dexter NM	48	B4
Diamond Bar CA	107	F3
Diamondhead MS	64	A3
Diamond Sprs. CA	31	F3
Diamondville WY	20	B4
Diaz AR	51	F2
D'Iberville MS	64	B3
Diboll TX	62	C3
Dickinson ND	12	C3
Dickinson TX	61	F1
Dickson OK	50	B3
Dickson TN	52	C2
Dierks AR	51	D4
Dighton KS	37	D3
Dilkon AZ	46	C2
Dilley TX	60	B2
Dillingham AK	74	B3
Dillon MT	10	B4
Dillon SC	54	C3
Dilworth MN	14	A3
Dimmitt TX	48	C3
Dinuba CA	44	B1
Dixfield ME	29	F1
Dixon CA	31	F3
Dixon IL	25	E4
Dixon MO	39	D4
Dodge Ctr. MN	24	B4
Dodge City KS	37	D4
Dodgeville WI	25	D3
Dolan Sprs. AZ	45	F2
Dolton IL	98	E6
Doña Ana NM	57	E1
Donaldsonville LA	63	F4
Donalsonville GA	65	E3
Doniphan MO	51	F1
Donna TX	60	C4
Dora AL	52	C4
Doraville GA	94	D1
Dormont PA	117	F3
Dorr MI	26	A3
Dos Palos CA	44	A1
Dothan AL	65	E3
Douglas AZ	56	C2
Douglas GA	65	E2
Douglas WY	21	E3
Douglas KS	37	F4
Douglasville GA	53	E4
Dousman WI	25	E3
Dover AR	51	E2
Dover DE	68	C4
Dover NH	29	F2

Dover NJ	69	D1
Dover OH	41	F1
Dover-Foxcroft ME	30	B3
Dowagiac MI	26	A4
Downers Grove IL	98	B5
Downey CA	44	C4
Downey ID	20	A3
Downingtown PA	68	C2
Downs KS	37	E2
Doylestown PA	69	D2
Draper UT	34	A1
Drayton ND	13	F1
Dresden TN	52	A1

Drew MS	52	A4
Drexel Hill PA	116	B3
Driggs ID	20	B2
Dripping Sprs. TX	59	F3
Druid Hills GA	94	C2
Drumright OK	50	B2
Dry Ridge KY	40	C3
Duarte CA	106	E2
Dublin CA	122	E3
Dublin GA	65	F1
Dublin OH	41	D2
Dublin TX	59	F2
Dublin VA	54	B1

Dubois ID	20	A1
DuBois PA	27	F4
Dubois WY	20	C2
Duluth GA	53	F4
Duluth MN	15	D3
Dumas AR	51	F4
Dumas TX	49	D2
Dulac LA	63	F4
Dulce NM	47	F1
Dumfries VA	42	C3
Duncan OK	50	A3
Duncanville TX	100	F3
Dundalk MD	68	B4
Dundee MI	26	C4
Dundee OR	8	A4

Dunedin FL	67	D2
Dunkirk IN	40	C1
Dunkirk NY	27	F3
Dunlap IN	26	A4
Dunlap TN	53	E2
Dunn NC	55	D2
Dunseith ND	13	D1
Du Quoin IL	39	F4
Durand MI	26	B3
Durand WI	24	C1
Durango CO	35	D4
Durant IA	25	D4

Entries in **bold color** indicate cities with detailed inset maps.

Durant—Fair Oaks Ranch **103**

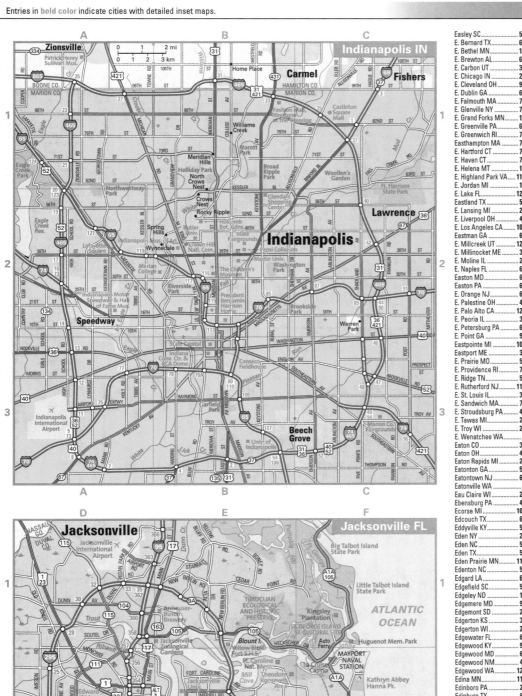

Indianapolis IN

Jacksonville FL

Durant MS 64 A1
Durant OK 50 B4
Durham CA 31 F2
Durham CT 70 C3
Durham NH 29 F2
Durham NC 55 D2

Duvall WA 8 B2
Dwight IL 39 F1
Dyer TN 52 B2
Dyersburg TN 52 B2
Dyersville IA 24 C3

E
Eagan MN 110 D4
Eagar AZ 47 D4
Eagle CO 35 E2
Eagle ID 19 D2
Eagle NE 38 A1

Eagle WI 25 E3
Eagle Butte SD 22 B1
Eagle Grove IA 24 B3
Eagle Lake MN 24 D1
Eagle Lake TX 61 E1
Eagle Mtn. UT 34 A1

Eagle Pass TX 60 A2
Eagle Pt. OR 17 E3
Earle AR 52 A2
Earlimart CA 44 B2
Earlington KY 40 A4
Early TX 59 E2

Easley SC 54 A3
E. Bernard TX 61 E1
E. Bethel MN 14 C4
E. Brewton AL 64 C3
E. Carbon UT 34 B2
E. Chicago IN 25 E4
E. Cleveland OH 99 F1
E. Dublin GA 65 F1
E. Falmouth MA 71 E3
E. Glenville NY 70 A1
E. Grand Forks MN 13 F2
E. Greenville PA 68 C1
E. Greenwich RI 71 D3
Easthampton MA 70 C2
E. Hartford CT 70 C2
E. Haven CT 70 C4
E. Helena MT 10 C3
E. Highland Park VA 119 B2
E. Jordan MI 16 B4
E. Lake FL 123 D1
Eastland TX 59 E1
E. Lansing MI 26 B3
E. Liverpool OH 41 F1
E. Los Angeles CA 106 C3
E. Millcreek UT 120 F2
E. Millinocket ME 30 B2
E. Moline IL 25 D4
E. Naples FL 67 E3
Easton MD 68 C4
Easton PA 69 D1
E. Orange NJ 69 E1
E. Palestine OH 41 F1
E. Palo Alto CA 122 C5
E. Peoria IL 39 F1
E. Petersburg PA 68 B2
E. Point GA 94 B4
Eastpointe MI 101 H1
Eastport ME 30 C2
E. Prairie MO 52 B1
E. Providence RI 71 D3
E. Ridge TN 53 E3
E. Rutherford NJ 112 B2
E. St. Louis IL 39 E3
E. Sandwich MA 71 F3
E. Stroudsburg PA 43 D1
E. Tawas MI 26 C1
E. Troy WI 25 E3
E. Wenatchee WA 8 C2

Eaton CO 35 F1
Eaton OH 40 C2
Eaton Rapids MI 26 B3
Eatonton GA 53 F4
Eatontown NJ 69 E2
Eatonville WA 8 B3
Eau Claire WI 24 C1
Ebensburg PA 42 B1
Ecorse MI 101 G3
Edcouch TX 60 C4
Eddyville KY 52 C1
Eden NY 27 F3
Eden NC 54 C1
Eden TX 59 E3
Eden Prairie MN 110 A4
Edenton NC 55 F2
Edgard LA 63 F4
Edgefield SC 54 A4
Edgeley ND 13 E3
Edgemere MD 95 D3
Edgemont SD 21 F2
Edgerton KS 38 B3
Edgerton WI 25 E3
Edgewater FL 67 E1
Edgewood KY 99 B3
Edgewood MD 68 B4
Edgewood NM 47 F3
Edgewood WA 123 B5
Edina MN 110 B3
Edinboro PA 27 E4
Edinburg TX 60 C4
Edinburgh IN 40 B2
Edison NJ 69 E1
Edmond OK 50 A2
Edmonds WA 8 B2
Edmonton KY 53 E1
Edna TX 61 E1
Edwards CO 35 E2
Edwards MS 63 E2
Edwardsville IL 39 E3
Effingham IL 39 F2
Egg Harbor City NJ 69 E2
Ehrenberg AZ 45 F4
Ekalaka MT 12 B4
Elba AL 65 D2
Elberton GA 54 A4
Elburn IL 25 E4
El Cajon CA 56 B4
El Campo TX 61 E1
El Cenizo TX 60 B3
El Centro CA 56 C4
El Cerrito CA 122 B2
Eldersburg MD 68 A3
Eldon MO 38 B4
Eldora IA 24 B3
El Dorado AR 51 E4
Eldorado IL 40 A4
El Dorado KS 37 F4
Eldorado TX 59 D3
El Dorado Sprs. MO 38 B4
El Dorado Springs MO 38 B4
Eldridge IA 25 D4
Eleanor WV 41 E3
Electra TX 49 F4
Eleele HI 72 B1

Elephant Butte NM 57 E1
Elfers FL 67 D1
Elgin IL 25 E4
Elgin ND 12 C3
Elgin OK 49 F3
Elgin OR 9 E4
Elgin SC 54 B3
Elgin TX 62 A3
Elizabeth CO 36 A2
Elizabeth NJ 69 E1
Elizabeth City NC 55 F1
Elizabethton TN 54 A1
Elizabethtown KY 40 B4
Elizabethtown NC 55 D3
Elizabethtown PA 68 B2
El Jebel CO 35 E2
Elk City OK 49 F2
Elk Grove CA 31 F3
Elk Grove Vil. IL 98 B3
Elkhart IN 26 A4
Elkhart KS 49 D1
Elkhorn NE 23 F4
Elkhorn WI 25 E3
Elkin NC 54 B1
Elkins AR 51 D2
Elkins WV 42 A3
Elko NV 33 D1
Elk Pt. SD 23 F3
Elkridge MD 68 B4
Elk Ridge UT 34 B2
Elk River MN 14 C4
Elkton KY 52 C1
Elkton MD 68 C3
Elkton SD 23 F1
Elkton VA 42 B3
Elkview WV 41 F3
Ellendale ND 13 E4
Ellensburg WA 8 C3
Ellenville NY 70 A1
Ellettsville IN 40 B2
Ellicott City MD 68 B4
Ellinwood KS 37 E3
Ellis KS 37 D3
Ellisville MS 64 B2
Ellsworth KS 37 E3
Ellsworth ME 30 B3
Ellsworth WI 24 C1
Ellwood City PA 41 F1
Elma WA 8 A3
Elm Creek NE 37 D1
Elmhurst IL 98 C4
Elmira NY 28 B4
Elmira Hts. NY 28 B4
Elmont NY 113 G4
El Monte CA 106 E2
Elmwood Park IL 98 C4
Elmwood Park NJ 112 B1
Elnora NY 70 A1
Eloy AZ 56 A1
El Paso IL 39 F1
El Paso TX 57 E2
El Reno OK 50 A2
El Rio CA 44 B3
Elroy WI 25 D2
Elsa TX 60 C4
Elsberry MO 39 E2
El Segundo CA 106 B3
Elwood IN 40 B1
Ely MN 15 D2
Ely NV 33 E2
Elyria OH 27 D4
Emerado ND 13 F2
Emerald Isle NC 55 E3
Emerson NE 23 E3
Eminence KY 40 C3
Emmaus PA 68 C1
Emmetsburg IA 24 A2
Emmett ID 19 D2
Emmonak AK 74 A2
Emory VA 54 B1
Empire LA 64 A4
Empire NV 32 B1
Emporia KS 38 A3
Emporia VA 55 E1
Emporium PA 27 F4
Encampment WY 21 E4
Encinitas CA 56 A3
Enderlin ND 13 E3
Endicott NY 28 B4
Endwell NY 28 B4
Enfield CT 29 E4
Enfield NC 29 E2
Enfield NC 55 E2
England AR 51 F3
Englewood CO 35 F2
Englewood FL 67 D3
Englewood NJ 112 D1
Englewood TN 53 E2
Enid OK 50 A1
Ennis MT 10 C4
Ennis TX 62 A1
Enoch UT 33 F4
Enosburg Falls VT 29 D1
Ensley FL 64 C3
Enterprise AL 65 D2
Enterprise NV 105 A2
Enterprise OR 9 E4
Enterprise UT 33 F4
Enumclaw WA 8 B3
Ephraim UT 34 B2
Ephrata PA 68 B2
Ephrata WA 9 D2

Epping NH 29 F3
Epworth IA 25 D3
Erath LA 63 E4
Erda UT 34 A1
Erie KS 38 A4
Erie PA 27 E4
Erlanger KY 99 A3
Erwin NC 55 D2
Erwin TN 54 A2
Escalon CA 31 F4
Escanaba MI 16 A4
Escobares TX 60 C4
Escondido CA 56 B3
Espanola NM 47 F2
Essex CT 70 D3
Essex MD 68 B3
Essex Jct. VT 29 D1
Essexville MI 26 C2
Estacada OR 8 B4
Estancia NM 47 F3
Estelle LA 111 C2
Estelline SD 23 E1
Ester AK 74 C2
Estero FL 67 E3
Estes Park CO 35 F1
Estherville IA 24 A2
Estill SC 66 B1
Estill Sprs. TN 53 D2
Ethete WY 20 C3
Etowah NC 54 A3
Etowah TN 53 E2
Euclid OH 27 D4
Eudora AR 63 F1
Eudora KS 38 B3
Eufaula AL 65 E2
Eufaula OK 50 C2
Eugene OR 17 E2
Euharlee GA 53 E4
Euless TX 100 E2
Eunice LA 63 E3
Eunice NM 58 B1
Eupora MS 52 B4
Eureka CA 31 D1
Eureka IL 39 F1
Eureka KS 38 A4
Eureka MO 39 E3
Eureka MT 10 A1
Eureka NV 33 D2
Eureka SD 13 D4
Eureka Sprs. AR 51 D1
Eustis FL 67 E1
Eutaw AL 64 C1
Evadale TX 62 C3
Evans CO 35 F1
Evans GA 54 A4
Eveleth MN 14 C2
Everett MA 96 E1
Everett WA 8 B2
Evergreen AL 64 C2
Evergreen CO 35 F2
Evergreen MT 10 A1
Evergreen Park IL 98 D5
Everson WA 8 B1
Ewa Beach HI 72 A1
Ewa Villages HI 72 A3
Excelsior Sprs. MO 38 B2
Exeter CA 44 B2
Exeter NH 71 E1
Experiment GA 53 F4
Exton PA 68 C2
Eyota MN 24 C2

F
Fabens TX 57 F2
Fairacres NM 57 E1
Fairbanks AK 74 C2
Fairborn OH 41 D2
Fairburn GA 53 E4
Fairbury IL 39 F1
Fairbury NE 37 F1
Fairfax CA 31 E4
Fairfax OK 50 B1
Fairfax SC 66 B1
Fairfax VA 68 A4
Fairfield CA 31 F3
Fairfield CT 70 B3
Fairfield IL 40 A3
Fairfield IA 39 D1
Fairfield ME 30 B3
Fairfield MT 10 C2
Fairfield OH 41 D2
Fairfield TX 62 B2
Fairfield Bay AR 51 E2
Fairfield Glade TN 53 E2
Fairhaven MA 71 E3
Fair Haven NJ 69 E2
Fair Haven VT 29 D2
Fairhope AL 64 B3
Fair Lawn NJ 112 B1
Fairmont MN 24 A2
Fairmont NC 55 D3
Fairmont WV 42 A2
Fairmount IN 40 C1
Fairmount NY 28 B3
Fairmount ND 14 A4
Fair Oaks Ranch TX 59 E4

Fairplains NC 54 B2
Fairport Harbor OH 27 D4
Fairview MT 12 B2
Fairview NJ 112 B2
Fairview NC 54 A2
Fairview OK 49 F1
Fairview TN 52 C2
Fairview UT 34 B2
Fairview Hts. IL 120 D2
Fairview Park OH 99 D2
Falfurrias TX 60 C3
Fall City WA 8 B2
Fallon NV 32 B2
Fall River MA 71 E3
Falls Church VA 124 B2
Falls City NE 38 A2
Fallston MD 68 B3
Falmouth KY 41 D3
Falmouth MA 71 E4
Falmouth VA 42 C3
Falmouth Foreside ME .. 29 F2
Fargo ND 14 A3
Faribault MN 24 B1
Farmers Branch TX 100 F1
Farmersville CA 44 B1
Farmersville TX 62 A1
Farmerville LA 63 E1
Farmington AR 51 D2
Farmington IL 39 E4
Farmington ME 29 F1
Farmington MI 101 E1
Farmington MN 24 B1

Farmington MS 52 B3
Farmington MO 39 E4
Farmington NH 29 F2
Farmington NM 47 E1
Farmington UT 34 A1
Farmington Hills MI .. 101 E1
Farmville NC 55 E2
Farmville VA 42 B4
Faulkton SD 23 D1
Fayette AL 52 C4
Fayette MS 63 F2
Fayette MO 38 C3
Fayetteville AR 51 D2
Fayetteville GA 53 F4
Fayetteville NC 55 D3
Fayetteville PA 68 A2
Fayetteville TN 53 D3
Fayetteville WV 41 F4
Federal Hts. CO 101 B1
Federalsburg MD 43 D3
Federal Way WA 8 B2
Fellsmere FL 67 F2
Fennimore WI 25 D3
Fenton MI 26 C3
Ferdinand IN 40 B3
Fergus Falls MN 14 A2
Ferguson MO 120 B1
Fernandina Beach FL .. 66 B3
Ferndale MD 95 B4
Ferndale MI 101 G1
Ferndale WA 8 B1
Fernley NV 32 B2

Ferriday LA 63 E2
Florala AL 65 D3
Ferris TX 62 A1
Ferron UT 34 B3
Ferrysburg MI 26 A2
Fessenden ND 13 E2
Festus MO 39 E3
Filer ID 19 E3
Fillmore CA 44 C3
Fillmore UT 34 A3
Findlay OH 41 D1
Finley ND 13 F2
Finley WA 9 D3
Finneytown OH 99 B2
Firebaugh CA 44 A1
Fishers IN 40 B2
Fishersville VA 42 B4
Fitchburg MA 71 D1
Fitzgerald GA 65 F2
Flagler Beach FL .. 66 B4
Flagstaff AZ 46 B3
Flandreau SD 23 F2
Flat Rock MI 26 C4
Flat Rock NC 54 A2
Flatwoods KY 41 E3
Fleetwood PA 68 C1
Flemingsburg KY 41 D3
Flemington NJ 69 D1
Fletcher NC 54 A2
Flint MI 26 C3
Flippin AR 51 E1
Flomaton AL 64 C3
Flora IL 39 F3
Flora IN 40 B1

Flora MS 63 F1
Floral City FL 67 D1
Floral Park NY 113 G3
Flora Vista NM 47 E1
Florence AL 52 C3
Florence AZ 56 B1
Florence CA 106 C3
Florence CO 35 F3
Florence KY 40 C3
Florence MT 10 A2
Florence OR 17 D2
Florence SC 54 C4
Floresville TX 60 C1
Florida NY 70 A3
Florida City FL 67 F4
Florin CA 31 F3
Florissant MO 120 B1
Flower Mound TX 100 E1
Floydada TX 49 D4
Flushing MI 26 C3
Foley AL 64 C4
Foley MN 14 C4
Folkston GA 66 B3
Folly Beach SC 66 C1
Folsom CA 31 F3
Fond du Lac WI 25 E2
Fontana CA 107 H2
Ford City CA 44 B3
Ford City PA 42 A1
Fords Prairie WA 8 B3
Fordyce AR 51 E4

Forest MS 64 A1
Forest VA 42 A4
Forest City IA 24 B2
Forest City NC 54 B2
Forestdale AL 96 A1
Forestdale MA 71 F3
Forest Grove OR 8 A4
Forest Hill TX 100 D3
Forest Lake MN 14 C4
Forest Park GA 53 F4
Forest Park IL 98 C4
Forest Park OH 99 B1
Forestville MD 124 E3
Forestville OH 99 C3
Forked River NJ 69 E3
Forks WA 8 A2
Forman ND 13 F3
Forrest City AR 51 F3
Forsyth GA 65 F1
Forsyth MO 51 E1
Forsyth MT 11 F3
Ft. Ashby WV 42 B2
Ft. Atkinson WI 25 E3
Ft. Belknap Agency MT .. 11 E1
Ft. Benton MT 11 D2
Ft. Bragg CA 31 D2
Ft. Branch IN 40 A3
Ft. Bridger WY 20 B4
Ft. Calhoun NE 23 F4
Ft. Collins CO 35 F1
Ft. Defiance AZ 47 D2
Ft. Dodge IA 24 A3

Ft. Edward NY 29 D2
Ft. Fairfield ME 30 C1
Ft. Gibson OK 50 C2
Ft. Hall ID 20 A3
Ft. Hancock TX 57 F2
Ft. Kent ME 30 B1
Ft. Lauderdale FL 67 F3
Ft. Lee NJ 69 E1
Ft. Lupton CO 35 F2
Ft. Madison IA 39 D1
Ft. Meade FL 67 E2
Ft. Mill SC 54 B3
Ft. Morgan CO 36 A1
Ft. Myers FL 67 E3
Ft. Myers Beach FL 67 E3
Ft. Myers Shores FL 67 E3
Ft. Myers Villas FL 67 E3
Ft. Oglethorpe GA 53 E3
Ft. Payne AL 53 E3
Ft. Pierce FL 67 F2
Ft. Pierre SD 22 C1
Ft. Scott KS 38 B4
Ft. Shawnee OH 41 D1
Ft. Smith AR 51 D2
Ft. Stockton TX 58 B3
Ft. Sumner NM 48 B3
Ft. Thomas KY 99 D3
Ft. Thompson SD 23 D2
Ft. Totten ND 13 E2
Ft. Valley GA 65 F1
Ft. Walton Beach FL 65 D3
Ft. Washakie WY 20 C3

Ft. Wayne IN 40 C1
Ft. Wingate NM 47 D2
Ft. Worth TX 62 A1
Ft. Yukon AK 74 C2
Fosston MN 14 A2
Foster City CA 122 C4
Fostoria OH 26 C4
Fountain CO 36 A3
Fountain Hills AZ 46 B4
Fountain Inn SC 54 A3
Fountain Valley CA 106 E5
Four Corners OR 17 C1
Fowler CA 44 B1
Fowler CO 36 A3
Fowler IN 40 A1
Fowlerville MI 26 B3
Foxboro MA 71 E2
Fox Lake IL 25 E3
Frackville PA 68 B1
Framingham MA 71 D2
Franconia VA 68 A4
Frankenmuth MI 26 C2
Frankfort IL 25 F4
Frankfort IN 40 B2
Frankfort KY 40 C3
Frankfort NY 28 C2
Franklin ID 20 A4
Franklin IN 40 B2
Franklin KY 53 D1
Franklin LA 71 D2
Franklin MA 71 D2
Franklin NE 37 E2
Franklin NH 29 E2
Franklin NJ 43 E1

Franklin NC 53 E2
Franklin OH 41 D2
Franklin PA 27 E4
Franklin TN 53 D2
Franklin TX 62 A1
Franklin VA 55 E1
Franklin WI 25 F3
Franklin Park IL 98 C3
Franklinton LA 64 A3
Fraser MI 101 H1
Frazer MT 11 F2
Frederick MD 68 A3
Frederick OK 49 F3
Fredericksburg TX 59 F3
Fredericksburg VA 42 C3
Fredericktown MO 39 E4
Fredonia AZ 46 B1
Fredonia KS 38 A4
Fredonia NY 27 F3
Fredonia WI 25 F2
Freeburg IL 39 E3
Freedom CA 31 D4
Freehold NJ 69 E2
Freeland MI 26 B2
Freeland PA 43 E1
Freeman SD 23 E2
Freeport IL 25 D3
Freeport ME 29 F2
Freeport NY 69 F1
Freeport TX 61 F1
Freer TX 60 C2
Fremont CA 31 D4

Fremont MI 26 A2
Fremont NE 23 F4
Fremont OH 26 C4
French Camp CA 31 F4
Frenchtown MT 10 A2
Fresno CA 44 B1
Friars Pt. MS 52 A3
Friday Harbor WA 8 B1
Fridley MN 110 C1
Friend NE 37 F1
Friendswood TX 61 F1
Friona TX 48 C3
Frisco CO 35 E2
Frisco TX 62 A1
Fritch TX 49 D2
Fromberg MT 11 E4
Frontenac KS 38 B4
Front Royal VA 42 B2
Frostburg MD 42 B2
Fruita CO 35 D3
Fruitland ID 18 C3
Fruitland MD 43 D3
Fruitland NM 47 E1
Fruitland Park FL 67 E1
Fryeburg ME 29 F2
Fullerton CA 44 C4
Fullerton NE 23 E4
Fulton IL 25 D4
Fulton KY 52 B1
Fulton MS 52 B3
Fulton MO 39 D3
Fulton NY 28 B2
Fulton TX 61 D2
Fuquay-Varina NC 55 D2

G
Gadsden AL 53 D3
Gaffney SC 54 B3
Gahanna OH 41 E2
Gainesville FL 66 A3
Gainesville GA 53 F3
Gainesville TX 50 B4
Gaithersburg MD 68 A4
Galax VA 54 B1
Galena AK 74 B2
Galena IL 25 D3
Galena KS 51 D1
Galesburg IL 39 E1
Galion OH 41 E1
Gallatin MO 38 C2
Gallatin TN 53 D1
Galliano LA 64 A4
Gallipolis OH 41 E3
Gallup NM 47 D2
Galt CA 31 F3
Galva IL 25 D4
Galveston TX 61 F1
Gambell AK 74 A2
Gamewell NC 54 B2
Ganado AZ 47 D2
Ganado TX 61 E1
Gardena CA 106 C3
Garden City GA 66 B2
Garden City KS 36 C4
Garden City MI 101 E1
Garden City MO 38 B3
Garden City SC 55 D4
Gardendale AL 53 D4
Garden Grove CA 106 E4
Garden Ridge TX 59 F4
Gardiner ME 29 F1
Gardiner MT 20 B1
Gardner KS 38 B3
Gardner MA 71 D1
Gardnerville NV 32 B2
Garfield NJ 112 B1
Garfield TX 62 A3
Garfield Hts. OH 99 F3
Garland TX 62 A1
Garland UT 20 A4
Garner IA 24 B3

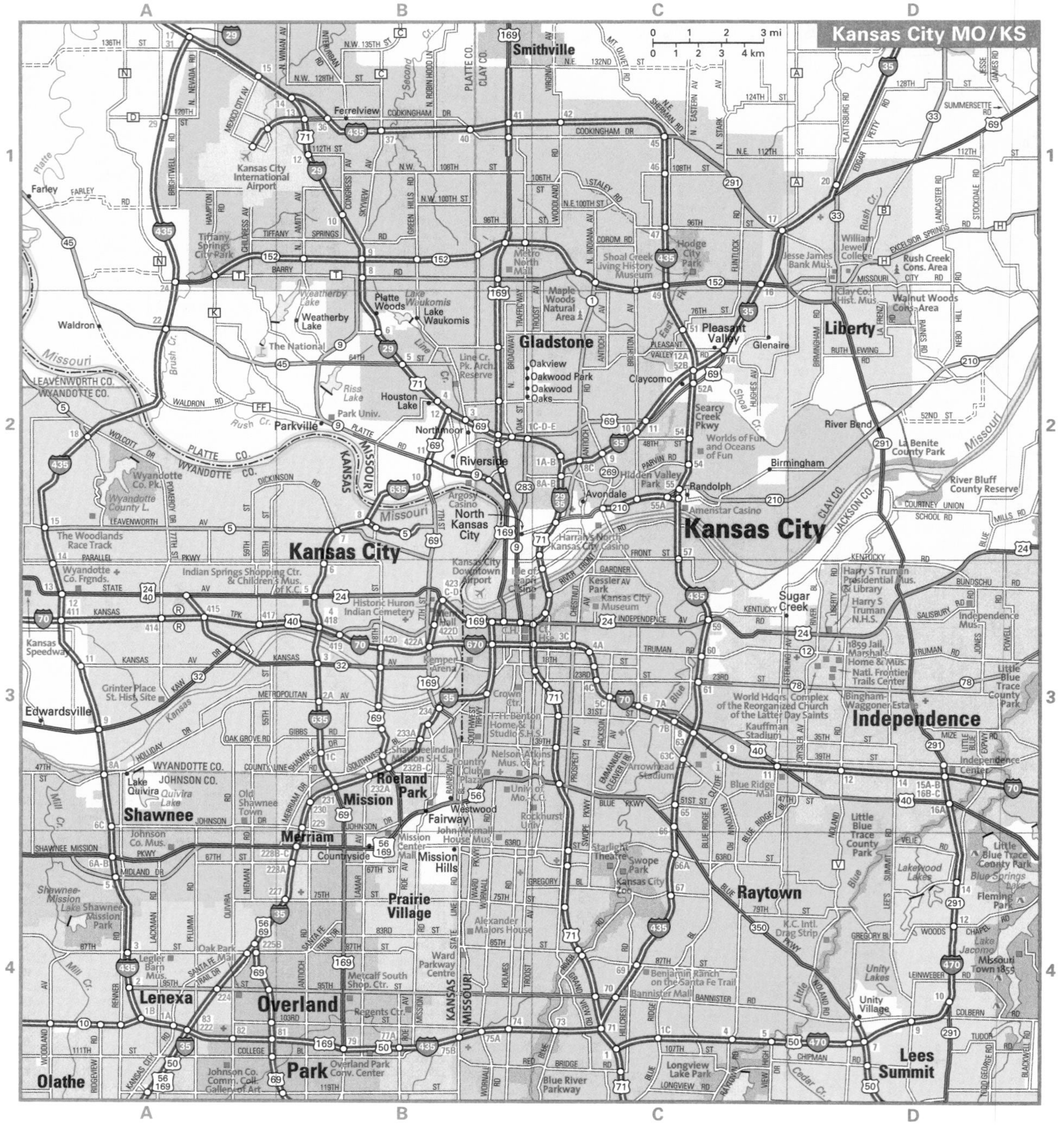

Kansas City MO/KS

Map labels include: Smithville, Ferrelview, Kansas City International Airport, Gladstone, Liberty, Parkville, Riverside, North Kansas City, Kansas City, Independence, Shawnee, Mission, Roeland Park, Merriam, Fairway, Prairie Village, Mission Hills, Overland Park, Lenexa, Olathe, Raytown, Lees Summit, Edwardsville.

Entries in **bold color** indicate cities with detailed inset maps.
Garner—Henagar 105

Garner NC 55 D2
Garnett KS 38 B3
Garretson SD 23 F2
Garrett IN 26 B4
Garrison ND 13 D2
Gary IN 25 F4
Gas City IN 40 C1
Gassville AR 51 E1
Gastonia NC 54 B2
Gate City VA 54 A1
Gatesville TX 59 F2
Gatlinburg TN 53 F2
Gautier MS 64 B3
Gaylord MI 26 B1
Gaylord MN 24 A1
Geary OK 49 F2
Genesee ID 9 E3
Geneseo IL 25 D4
Geneseo NY 28 A3
Geneva AL 65 D3
Geneva IL 25 E4
Geneva NE 37 F1
Geneva NY 28 B3
Geneva OH 27 D4
Geneva WA 8 B1
Genoa IL 25 E4
Genoa NE 23 E4
Genoa City WI 25 E3
Gentry AR 51 D1
Georgetown CO 35 F2
Georgetown DE 43 D3
Georgetown ID 20 A3
Georgetown IL 40 A2
Georgetown IN 40 B3
Georgetown KY 40 C3
Georgetown MA 71 E1
Georgetown OH 41 D3
Georgetown SC 55 D4
Georgetown TX 59 F3
George West TX 60 C2
Georgiana AL 64 C2
Gering NE 22 A4
Gerlach NV 32 B1
Germantown MD 68 A4
Germantown OH 41 D2
Germantown TN 52 A3
Germantown WI 25 E2
Gettysburg PA 68 A3
Gettysburg SD 22 C1
Gibbon NE 37 E1
Gibsonburg OH 26 C4
Gibson City IL 39 F1
Gibsonton FL 67 D2
Giddings TX 62 A4
Gifford FL 67 F2
Gig Harbor WA 8 B2
Gila Bend AZ 57 E3
Gilbert AZ 46 B4
Gilbert MN 15 D2
Gilcrest CO 35 F1
Gillespie IL 39 E3
Gillette WY 21 E1
Gilmer TX 62 C1
Gilroy CA 44 A1
Girard KS 38 B4
Girard PA 27 E4
Gladewater TX 62 C1
Gladstone MI 16 A4
Gladstone MO 104 C2
Gladstone OR 118 B3
Gladwin MI 26 B2
Glasgow DE 68 C3
Glasgow KY 53 D1
Glasgow MT 11 F1
Glassboro NJ 69 D3
Glassmanor MD 124 D3
Glastonbury CT 70 C3
Glen Allen VA 119 A1
Glen Burnie MD 68 B4
Glencoe AL 53 D4
Glencoe MN 24 A1
Glen Cove NY 69 F1
Glendale AZ 46 B4
Glendale CA 44 C4
Glendale WI 109 D1
Glendale Hts. IL 98 B4
Glendive MT 12 B2
Glendora CA 107 F2
Glen Ellyn IL 98 B4
Glenmora LA 63 E3
Glennallen AK 74 C3
Glenns Ferry ID 19 E3
Glennville GA 66 A2
Glenpool OK 50 C2
Glenrock WY 21 E3
Glen Rose TX 59 F1
Glens Falls NY 29 D2
Glen Ullin ND 12 C3
Glenview IL 98 C2
Glenville WV 41 F3
Glenwood AR 51 D3
Glenwood IA 38 A1
Glenwood MN 14 B4
Glenwood Sprs. CO 35 E2
Glide OR 17 E2
Globe AZ 46 C4
Glorieta NM 48 A2
Gloucester MA 71 E1
Gloucester VA 42 C4
Gloucester City NJ 116 D4
Gloucester Pt. VA 42 C4
Gloversville NY 28 C3
Goddard KS 37 F4
Godfrey IL 39 E3
Goffstown NH 29 E3
Gold Bar WA 8 B2
Gold Beach OR 17 D3
Golden CO 35 F2
Goldendale WA 8 C4
Golden Gate FL 67 E3
Golden Meadow LA 64 A4
Golden Valley MN 110 B2
Goldsboro NC 55 D2
Goldsby OK 50 A3
Goldthwaite TX 59 F2
Goleta CA 44 B3
Goliad TX 61 D2
Gonzales CA 44 A1
Gonzales LA 63 F3
Gonzales TX 61 D1
Gonzalez FL 64 C3
Good Hope AL 53 D4
Gooding ID 19 E3
Goodland KS 36 C2
Goodlettsville TN 53 D1
Goodman MS 64 A1
Goodview MN 24 C2
Goodwater AL 65 D1
Goodwell OK 49 D1
Goodyear AZ 46 B4
Goose Creek SC 66 C1
Gordo AL 52 C4
Gordon GA 65 F1
Gordon NE 22 B3
Gordonsville VA 42 B4
Gorham NH 29 F1
Goshen IN 26 A4
Goshen NY 70 A3
Gosnell AR 52 A2
Gothenburg NE 37 D1
Gould AR 51 F4
Goulds FL 67 F4
Gouverneur NY 28 C1
Gowanda NY 27 F3
Grace ID 20 A3
Grafton ND 13 F1
Grafton WV 42 A2
Grafton WI 25 F2
Graham NC 54 C2
Graham TX 59 F1
Grambling LA 63 E1
Gramercy LA 63 F4
Granbury TX 59 F1
Granby CO 35 F2
Granby MO 51 D1
Grand Bay AL 64 B3
Grand Blanc MI 26 C3
Grand Canyon AZ 46 B2
Grandfield OK 49 F3
Grand Forks ND 13 F2
Grand Haven MI 26 A3
Grand Island NE 37 E1
Grand Isle LA 64 A4
Grand Jct. CO 35 D3
Grand Ledge MI 26 B3
Grand Prairie TX 100 E3
Grand Rapids MI 26 A3
Grand Rapids MN 14 C2
Grand Saline TX 62 B1
Grandview WA 8 C3
Grandview Plaza KS 37 F3
Granger WA 8 C3
Grangeville ID 9 F4
Granite OK 49 F3
Granite City IL 39 E3
Granite Falls MN 23 F1
Granite Falls NC 54 B2
Granite Falls WA 8 B2
Granite Quarry NC 54 C2
Granite Shoals TX 59 F3
Graniteville SC 54 B4
Grant NE 36 C1
Grants NM 47 E3
Grantsdale MT 10 A3
Grants Pass OR 17 E3
Grantsville UT 34 A1
Granville NY 29 D2
Granville OH 41 E2
Grapeland TX 62 B2
Grapevine TX 62 A1
Grass Valley CA 31 F2
Gravette AR 51 D1
Gray LA 63 F4
Grayslake IL 98 B1
Grayson KY 41 E3
Great Bend KS 37 E3
Great Falls MT 10 C2
Great Falls SC 54 B3
Great Neck NY 69 F1
Greece NY 28 A2
Greeley CO 35 F1
Green OR 17 E2
Greenacres CA 44 B2
Greenacres FL 67 F3
Green Bay WI 25 E1
Greenbelt MD 68 A4
Greenbrier AR 51 E2
Greenbrier TN 53 D1
Greencastle IN 40 B2
Greencastle PA 42 B2
Green Cove Sprs. FL 66 B4
Greendale WI 109 C3
Greeneville TN 54 A2
Greenfield CA 44 A1
Greenfield IN 40 C2
Greenfield IA 24 A4
Greenfield MA 70 C1
Greenfield OH 41 D2
Greenfield TN 52 B2
Greenfield WI 109 C3
Green Forest AR 51 D1
Green River WY 20 C4
Greensboro AL 64 C1
Greensboro GA 53 F4
Greensboro NC 54 C2
Greensburg IN 40 C2
Greensburg KS 37 E4
Greensburg KY 40 C4
Greensburg PA 42 A1
Greentown IN 40 B1
Green Valley AZ 56 B2
Green Valley MD 68 A3
Greenville AL 65 D2
Greenville IL 39 F3
Greenville KY 40 A4
Greenville ME 30 B3
Greenville MI 26 B2
Greenville MS 51 F4
Greenville NC 55 E2
Greenville OH 40 C2
Greenville PA 27 E4
Greenville RI 71 D3
Greenville SC 54 A3
Greenville TX 62 B1
Greenwich CT 70 B4
Greenwood AR 51 D3
Greenwood IN 40 B2
Greenwood LA 63 D1
Greenwood MS 52 A4
Greenwood SC 54 A3
Greenwood Lake NY 70 A4
Greenwood Vil. CO 101 C4
Greer SC 54 A3
Gregory SD 23 D2
Gregory TX 61 D2
Grenada MS 52 A4
Gresham OR 8 B4
Gresham Park GA 94 C3
Gretna LA 64 A4
Gretna NE 23 F4
Greybull WY 21 D1
Gridley CA 31 F2
Griffin GA 53 F4
Grifton NC 55 E2
Grimes IA 24 B4
Grinnell IA 24 B4
Groesbeck TX 62 B2
Grosse Pointe Farms MI 101 H2
Grosse Pointe Park MI 101 H2
Grosse Pointe Woods MI 101 H2
Groton CT 71 D3
Groton SD 13 F4
Grottoes VA 42 B3
Grove OK 50 C1
Grove City OH 41 D2
Grove City PA 27 E4
Groveland CA 32 A4
Groveland MA 71 E1
Grover Beach CA 44 A3
Groves TX 63 D4
Grovetown GA 54 A4
Gruetli-Laager TN 53 E2
Grundy Ctr. IA 24 B3
Guadalupe CA 44 A3
Gueydan LA 63 E4
Guilford CT 70 C4
Guernsey WY 21 F3
Guin AL 52 C4
Gulf Breeze FL 64 C3
Gulfport MS 64 B3
Gulf Shores AL 64 C3
Gun Barrel City TX 62 B1
Gunnison CO 35 E3
Gunnison UT 34 A2
Guntersville AL 53 D3
Guntown MS 52 B3
Gurdon AR 51 E4
Gustine CA 31 F4
Guthrie KY 52 C1
Guthrie OK 50 A2
Guthrie Ctr. IA 24 A4
Guttenberg IA 24 C3
Guttenberg NJ 112 D2
Guymon OK 49 D1
Gwinner ND 13 F3
Gypsum CO 35 E2

H
Hacienda Hts. CA 106 E3
Hackberry LA 63 D4
Hackensack NJ 69 E1
Hackettstown NJ 69 D1
Hackleburg AL 52 C3
Haddonfield NJ 116 D4
Hagerman ID 19 E3
Hagerman NM 58 A1
Hagerstown MD 42 B2
Hahnville LA 63 F4
Haiku HI 73 D1
Hailey ID 19 E2
Haines AK 75 D3
Haines City FL 67 E2
Halawa HI 72 C3
Hale Ctr. TX 49 D4
Haleiwa HI 72 A2
Haleyville AL 52 C3
Half Moon NC 55 E3
Half Moon Bay CA 31 E4
Hallandale Beach FL 67 F4
Hallettsville TX 61 D1
Hallowell ME 29 F1
Halls TN 52 B2
Halls Crossroads TN 53 F2
Hallsville TX 62 C1
Halstead KS 37 F4
Haltom City TX 100 D2
Hamburg AR 51 F4
Hamburg NY 27 F3
Hamburg PA 68 C1
Hamden CT 70 C3
Hamilton AL 52 C4
Hamilton IL 39 D1
Hamilton MO 38 B2
Hamilton MT 10 A3
Hamilton NY 28 C3
Hamilton OH 40 C2
Hamilton TX 59 F2
Ham Lake MN 14 C4
Hamlet NC 54 C3
Hamlin TX 59 E1
Hamlin WV 41 E3
Hammond IN 25 F4
Hammond LA 63 F3
Hammonton NJ 69 D3
Hampden ME 30 B3
Hampshire IL 25 E4
Hampstead MD 68 B3
Hampton AR 51 E4
Hampton GA 53 F4
Hampton IA 24 B3
Hampton NH 71 E1
Hampton SC 66 B1
Hampton VA 55 F1
Hampton Bays NY 70 C4
Hampton Beach NH 71 E1
Hamtramck MI 101 G2
Hanahan SC 66 C1
Hanamaulu HI 72 B1
Hanapepe HI 72 B1
Hanceville AL 53 D4
Hancock MI 15 F2
Hanford CA 44 B1
Hankinson ND 14 A3
Hanna WY 21 E4
Hannibal MO 39 D2
Hanover IN 40 C3
Hanover NH 29 E2
Hanover PA 68 A3
Hanover Park IL 98 A3
Hansen ID 19 E3
Harahan LA 111 B2
Harbor OR 17 D4
Hardeeville SC 66 B1
Hardin MT 11 F4
Hardinsburg KY 40 B4
Hardwick VT 29 E1
Harker Hts. TX 59 F3
Harlan IA 23 F4
Harlan KY 53 F1
Harlem MT 11 E1
Harlingen TX 61 D4
Harlowton MT 11 D3
Harold KY 41 E4
Harper KS 37 E4
Harpersville AL 53 D4
Harper Woods MI 101 H2
Harrah OK 50 B2
Harriman TN 53 E2
Harrington DE 68 C4
Harrisburg AR 51 F2
Harrisburg IL 39 F4
Harrisburg NC 54 B2
Harrisburg OR 17 E1
Harrisburg PA 68 B2
Harrisburg SD 23 E2
Harrison AR 51 E1
Harrison MI 26 B2
Harrison NJ 112 B3
Harrison TN 53 E3
Harrisonburg VA 42 B3
Harrisonville MO 38 B3
Harrisville WV 41 F2
Harrodsburg KY 40 C4
Harrogate TN 53 F1
Hartford AL 65 D3
Hartford CT 70 C3
Hartford KY 40 B4
Hartford MI 26 A3
Hartford SD 23 E2
Hartford WI 25 E2
Hartford City IN 40 C1
Hartington NE 23 E3
Hartley IA 23 F2
Harts WV 41 E4
Hartselle AL 53 D3
Hartshorne OK 50 C3
Hartsville SC 54 C3
Hartsville TN 53 D1
Hartwell GA 54 A3
Harvard IL 25 E3
Harvard NE 37 E1
Harvest AL 53 D3
Harvey IL 98 D6
Harvey LA 111 C2
Harvey ND 13 E2
Harveys Lake PA 28 B4
Harwinton CT 70 B3
Harwood ND 14 A3
Hasbrouck Hts. NJ 112 C1
Haskell AR 51 E3
Haskell OK 50 C2
Haskell TX 59 E1
Hastings MI 26 B3
Hastings MN 24 B1
Hastings NE 37 E1
Hatch NM 57 E1
Hattiesburg MS 64 A2
Hatton ND 13 F2
Haughton LA 63 D1
Hauula HI 72 A2
Havana IL 39 E1
Havelock NC 55 E3
Haven KS 37 F4
Haverhill MA 71 E1
Haverstraw NY 70 A4
Havre MT 11 D1
Havre de Grace MD 68 B3
Hawarden IA 23 F3
Hawkinsville GA 65 F1
Hawley MN 14 A3
Hawthorne CA 106 C3
Hawthorne NV 32 B3
Hayden CO 35 E1
Hayden ID 9 E2
Hayesville OR 17 E1
Haynesville LA 63 D1
Hays KS 37 E3
Hays MT 11 E2
Haysville KS 37 F4
Hayti MO 52 B1
Hayward CA 31 F4
Hayward WI 15 D4
Hazard KY 41 D4
Hazel Green AL 53 D3
Hazel Park MI 101 G1
Hazelton ID 19 E3
Hazelwood MO 120 A1
Hazen AR 51 F3
Hazen ND 13 D2
Hazlehurst GA 66 A2
Hazlehurst MS 63 F2
Hazleton PA 68 C1
Headland AL 65 E2
Healdsburg CA 31 E3
Healdton OK 50 A3
Healy AK 74 C2
Hearne TX 62 A3
Heart Butte MT 10 B1
Heath OH 41 E2
Heavener OK 51 D3
Hebbronville TX 60 C3
Heber AZ 46 C3
Heber City UT 34 B1
Heber Sprs. AR 51 E2
Hebron IN 25 F4
Hebron NE 37 F1
Hebron ND 12 C3
Heeia HI 72 A3
Heflin AL 53 E4
Helena AL 53 D4
Helena AR 52 A3
Helena GA 65 F2
Helena MT 10 C3
Hellertown PA 68 C1
Helotes TX 59 F4
Helper UT 34 B2
Hemet CA 45 D4
Hemingford NE 22 A3
Hempstead NY 69 F1
Hempstead TX 62 B4
Henagar AL 53 D3

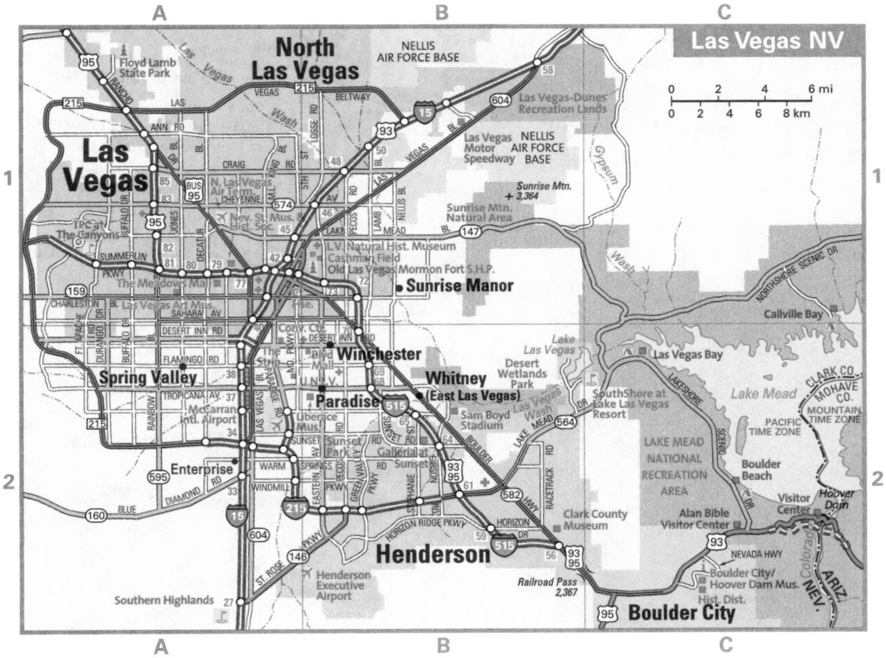

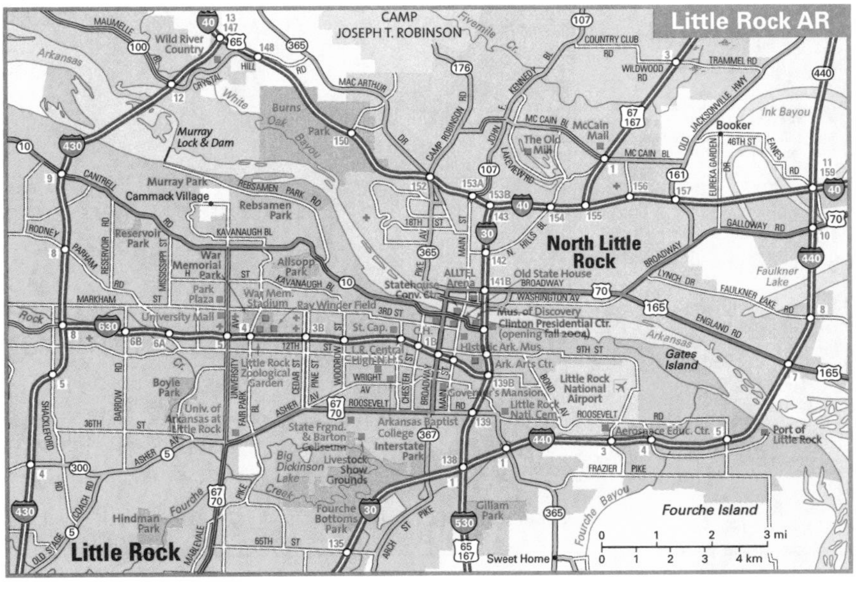

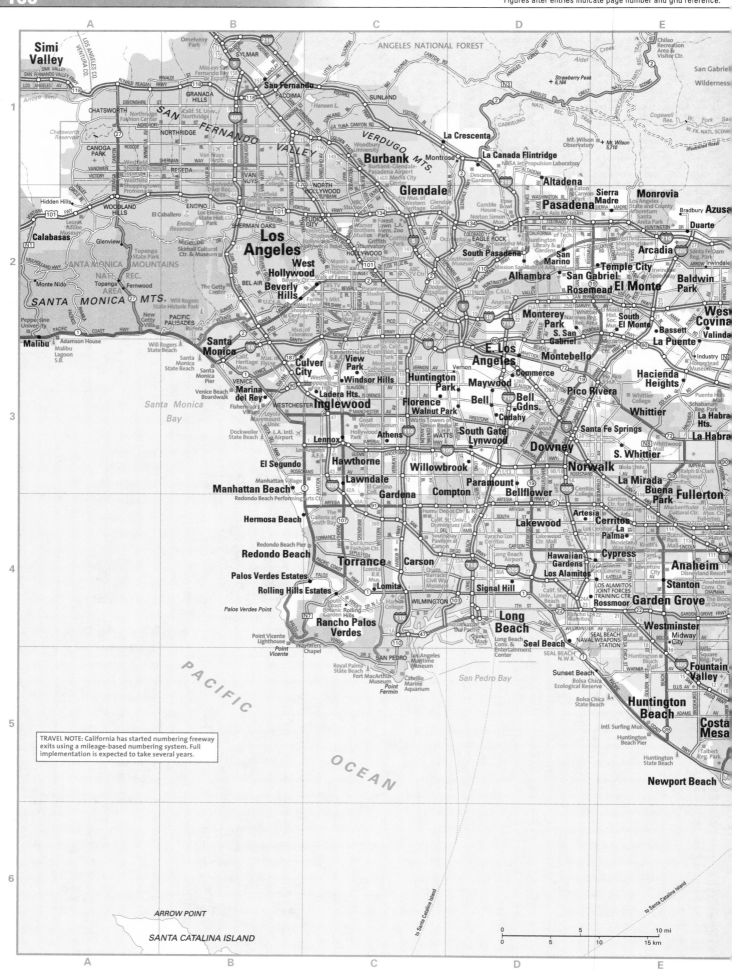

TRAVEL NOTE: California has started numbering freeway exits using a mileage-based numbering system. Full implementation is expected to take several years.

Entries in **bold color** indicate cities with detailed inset maps.

107

Los Angeles CA

Figures after entries indicate page number and grid reference.

Henderson KY 40 A4
Henderson LA 63 E3
Henderson NE 37 F1
Henderson NV 45 E2
Henderson NC 55 D1
Henderson TN 52 B2
Henderson TX 62 C2
Hendersonville NC 54 A2
Hendersonville TN 53 D1
Hennessey OK 50 A2
Henniker NH 29 E2
Henrietta TX 49 F4
Henry IL 39 F1
Henryetta OK 50 C2
Hephzibah GA 54 A4
Heppner OR 9 D4
Hercules CA 122 C1
Hereford TX 48 C3
Herington KS 37 F3
Herkimer NY 28 C3
Hermann MO 39 D3
Hermantown MN 15 D3
Hermiston OR 9 D4
Hermitage PA 27 E4
Hermosa Beach CA 106 B4
Hernandez NM 47 F2
Hernando FL 67 D1
Hernando MS 52 A3
Herndon VA 68 A4
Herrin IL 39 F4
Hershey PA 68 B2
Hertford NC 55 F2
Hesperia CA 45 D3
Hesston KS 37 F3
Hettinger ND 12 C4
Hewitt TX 62 A2
Heyburn ID 19 F3
Hialeah FL 67 F4
Hialeah Gardens FL 109 A3
Hiawatha IA 24 C4
Hiawatha KS 38 A2
Hibbing MN 14 C2
Hickman KY 52 B1
Hickman NE 37 F1
Hickory NC 54 B2
Hicksville NY 69 F1
Hicksville OH 26 B4
Hidalgo TX 60 C4
Higginsville MO 38 C3
High Bridge NJ 69 D1
Highland CA 107 K2
Highland IL 39 F3
Highland NY 70 A3
Highland Falls NY 70 A3
Highland Park IL 25 F3
Highland Park MI 101 G2
Highlands NJ 69 E2
Highland Sprs. VA 119 B2
Highlands Ranch CO 101 B4
Highmore SD 23 D1
High Pt. NC 54 C2
High Sprs. FL 66 A4
Hightstown NJ 69 E2
Hildale UT 46 A1
Hill City KS 37 D2
Hill City SD 22 A2
Hillcrest Hts. MD 124 D3
Hilliard OH 100 A2
Hillsboro IL 39 F2
Hillsboro KS 37 F3
Hillsboro MO 39 E3
Hillsboro ND 13 F2
Hillsboro OH 41 D2
Hillsboro OR 8 B4
Hillsboro TX 62 A2
Hillsborough NH 29 E3
Hillsborough NC 55 D2
Hillsdale MI 26 B4
Hillside NJ 112 A4
Hillsville VA 54 B1
Hillview KY 40 C4
Hilmar CA 31 F4
Hilo HI 73 F3
Hilton NY 28 A2
Hilton Head Island SC 66 B2
Hines OR 18 B2
Hinesburg VT 29 D1
Hinesville GA 66 B2
Hinsdale IL 98 C5
Hinsdale NH 70 C1
Hinton OK 49 F2
Hinton WV 41 F4
Hobart OK 49 F3
Hobbs NM 58 B1
Hobe Sound FL 67 F3
Hoboken NJ 112 C3
Hockessin DE 68 C3
Hodgenville KY 40 C4
Hoffman Estates IL 98 B3
Hogansville GA 65 E1
Hohenwald TN 52 C2
Hoisington KS 37 E3
Hokes Bluff AL 53 D4
Holbrook AZ 46 C3
Holbrook MA 71 E2
Holbrook NY 69 F1
Holcomb KS 36 C4
Holden MA 71 D2
Holden MO 38 B3
Holdenville OK 50 B3
Holdrege NE 37 D1
Holiday FL 67 D1
Holladay UT 120 F3
Holland MI 26 A3
Hollandale MS 51 F4
Holliday TX 49 F4
Hollidaysburg PA 42 B1
Hollins VA 42 A4
Hollis NH 71 D2
Hollis OK 49 E3
Hollister CA 44 A1
Hollister MO 51 E1
Holliston MA 71 D2
Holly CO 36 C4
Holly MI 26 C3
Holly Hill FL 66 B4
Holly Sprs. GA 53 E4
Holly Sprs. MS 52 B3
Holly Sprs. NC 55 D2
Hollywood FL 67 F4
Hollywood SC 66 C1
Hollywood Park TX 59 F4
Holmen WI 24 C2
Holmes Beach FL 67 D2
Holstein IA 23 F3
Holt AL 52 C4
Holt MI 26 B3
Holton KS 38 A2
Holts Summit MO 39 D3
Holtville CA 56 C4
Holualoa HI 73 E3
Holyoke CO 36 B1
Holyoke MA 70 C2
Homedale ID 18 C2
Homer AK 74 C3
Homer LA 63 D1
Homer NY 28 B3
Homer Glen IL 98 C6
Homerville GA 66 A3
Homestead FL 67 F4
Homewood AL 96 B2
Homewood IL 98 D6
Hominy OK 50 B1
Homosassa Sprs. FL 67 D1
Honalo HI 73 E3
Honaunau HI 73 E3
Hondo TX 59 E4
Honea Path SC 54 A3
Honesdale PA 28 C4
Honey Grove TX 50 C4
Honeyville UT 20 A4
Honokaa HI 73 F2
Honokowai HI 73 D1
Honolulu HI 72 A3
Hood River OR 8 B4
Hooker OK 49 D1
Hooks TX 51 D4
Hoonah AK 75 D4
Hooper NE 23 F4
Hooper Bay AK 74 A3
Hoopeston IL 40 A1
Hoosick Falls NY 70 B1
Hoover AL 53 D4
Hopatcong NJ 69 D1
Hope AR 51 D4
Hope IN 40 C2
Hope Mills NC 55 D3
Hopewell TN 53 E2
Hopewell VA 42 C4
Hopkins MN 110 A3
Hopkinsville KY 52 C1
Hoquiam WA 8 A3
Horace ND 14 A3
Horicon WI 25 E2
Horizon City TX 57 F2
Hornell NY 28 A3
Horn Lake MS 52 A3
Horse Cave KY 40 B4
Horseheads NY 28 B3
Horse Pasture VA 54 C1
Horseshoe Bend AR 51 F1
Horseshoe Bend ID 19 D2
Horsham PA 69 D2
Horton KS 38 A2
Hortonville WI 25 E1
Hot Sprs. AR 51 E3
Hot Sprs. MT 10 A2
Hot Sprs. SD 22 A2
Hot Sprs. Vil. AR 51 E3
Houck AZ 47 D2
Houghton MI 15 F2
Houghton Lake MI 26 B1
Houlton ME 30 C2
Houma LA 63 F4
Houston AK 74 C3
Houston MS 52 B4
Houston MO 39 D4
Houston TX 61 F1
Hoven SD 13 E4
Howard SD 23 E2
Howard WI 25 E1
Howard Lake MN 24 A1
Howards Grove WI 25 F2
Howe TX 50 B4
Howell MI 26 B3
Howland ME 30 B3
Hoxie AR 51 F2
Hoxie KS 37 D2
Hoyt Lakes MN 15 D2
Huachuca City AZ 56 B2
Hubbard OH 27 E4
Hubbard OR 8 B4
Hubbard TX 62 A2
Hudson CO 36 A2
Hudson FL 67 D1
Hudson IA 24 C3
Hudson MA 71 D2
Hudson MI 26 B4
Hudson NH 71 D1
Hudson NY 70 A2
Hudson NC 54 B2
Hudson TX 62 C2
Hudson WI 24 B1
Hudson WY 20 C3
Hudson Falls NY 29 D2
Hudsonville MI 26 A3
Hueytown AL 53 D4
Hughes AR 52 A3
Hughes Sprs. TX 62 C1
Hughson CA 31 F4
Hugo MN 14 C4
Hugo OK 50 C4
Hugoton KS 36 C4
Hulett WY 21 F1
Hull IA 23 F2
Hull MA 71 E2
Humble TX 62 C4
Humboldt AZ 46 B3
Humboldt IA 24 A3
Humboldt KS 38 A4
Humboldt NE 38 A1
Humboldt NV 32 C1
Humboldt SD 23 E2
Humboldt TN 52 B2
Hummelstown PA 68 B2
Hungry Horse MT 10 A1
Huntersville NC 54 B2
Huntingburg IN 40 B3
Huntingdon PA 42 B1
Huntingdon TN 52 B2
Huntington IN 40 C1
Huntington NY 69 F1
Huntington TX 62 C2
Huntington UT 34 B2
Huntington WV 41 E3
Huntington Beach CA 44 C4
Huntington Park CA 106 D3
Huntley IL 25 E3
Huntley MT 11 E4
Huntsville AL 53 D3
Huntsville AR 51 D2
Huntsville MO 39 D2
Huntsville TX 62 B3
Hurley NM 57 D1
Hurley NY 70 A2
Hurley WI 15 E3
Huron CA 44 B1
Huron OH 26 C4
Huron SD 23 D1
Hurricane UT 33 F4
Hurricane WV 41 E3
Hurst TX 100 D2
Hutchinson KS 37 F4
Hutchinson MN 24 A1
Huxley IA 24 B4
Hyannis MA 71 F3
Hyattsville MD 124 D1
Hyde Park NY 70 A3
Hyrum UT 20 A4

I

Icard NC 54 B2
Idabel OK 50 C4
Ida Grove IA 23 F3
Idaho Falls ID 20 A2
Idaho Sprs. CO 35 F2
Idalou TX 49 D4
Idyllwild CA 45 D4
Idylwood VA 124 A2
Ilion NY 28 C3
Imlay City MI 26 C3
Immokalee FL 67 E3
Imperial CA 56 C4
Imperial NE 36 C1
Imperial Beach CA 56 B4
Incline Vil. NV 32 A2
Independence IA 24 C3
Independence KS 38 A4
Independence KY 40 C3
Independence LA 63 F3
Independence MO 38 B3
Independence OR 17 E1
Indiana PA 42 A1
Indianapolis IN 40 B2
Indian Harbour Beach FL 67 F2
Indian Head MD 42 C3
Indianola IA 24 B4
Indianola MS 52 A4
Indian River MI 16 C4
Indian River Estates FL 67 F2
Indian River Shores FL 67 F2
Indian Rocks Beach FL 67 D2
Indian Sprs. NV 45 E1
Indiantown FL 67 F3
Indian Trail NC 54 B3
Indian Wells CA 45 E4
Indio CA 45 E4
Inez TX 61 D1
Ingleside TX 61 D2
Inglewood CA 44 C4
Ingram TX 59 E4
Inkom ID 20 A3
Inkster MI 101 F3
Inman KS 37 F3
Inman SC 54 A3
Inola OK 50 C2
International Falls MN 14 C1
Inver Grove Hts. MN 110 D4
Inverness FL 67 D1
Inverness MS 52 A4
Inwood NY 113 G5
Inwood WV 42 B2
Iola KS 38 A4
Iona ID 20 A2
Ione CA 31 F3
Ionia MI 26 B3
Iowa LA 63 D3
Iowa City IA 24 C4
Iowa Falls IA 24 B3
Iowa Park TX 49 F4
Ipswich MA 71 E1
Ipswich SD 13 E4
Irmo SC 54 B4
Irondale AL 96 B1
Irondequoit NY 28 A2
Iron Mtn. MI 15 F4
Iron River MI 15 F3
Ironton MO 39 E4
Ironton OH 41 E3
Ironwood MI 15 E3
Irrigon OR 9 D4
Irvine CA 44 C4
Irvine KY 41 D4
Irving TX 62 A1
Irvington NJ 112 A3
Isanti MN 14 C4
Ishpeming MI 16 A3
Islamorada FL 67 E4
Island Pond VT 29 E1
Isla Vista CA 44 B3
Isle of Hope GA 66 B2
Isle of Palms SC 66 C1
Islip NY 69 F1
Italy TX 62 A2
Itasca TX 62 A2
Ithaca MI 26 B2
Ithaca NY 28 B3
Itta Bena MS 52 A4
Iuka MS 52 C3
Ivanhoe CA 44 B1
Ivins UT 33 F4

J

Jackpot NV 19 E4
Jacksboro TN 53 F1
Jacksboro TX 50 A4
Jackson AL 64 C2
Jackson CA 32 A3
Jackson GA 53 F4
Jackson KY 41 D4
Jackson LA 63 F3
Jackson MI 26 B3
Jackson MN 24 A2
Jackson MS 63 F1
Jackson MO 39 F4
Jackson OH 41 E3
Jackson SC 54 B4
Jackson TN 52 B2
Jackson WI 25 E2
Jackson WY 20 B2
Jacksonville AL 53 D4
Jacksonville AR 51 E3
Jacksonville FL 66 B3
Jacksonville IL 39 E2
Jacksonville NC 55 E3
Jacksonville OR 17 E3
Jacksonville TX 62 C2
Jacksonville Beach FL 66 B3
Jaffrey NH 71 D1
Jal NM 58 B2
Jamesburg NJ 69 E2
Jamestown CA 32 A4
Jamestown KY 53 E1
Jamestown NY 27 F4
Jamestown ND 13 E3
Jamestown RI 71 D3
Jamestown TN 53 E1
James Town WY 20 C4
Jamul CA 56 B4
Janesville MN 24 B2
Janesville WI 25 E3
Jarales NM 47 F3
Jarrettsville MD 68 B3
Jasmine Estates FL 67 D1
Jasonville IN 40 A2
Jasper AL 52 C4
Jasper GA 53 F3
Jasper IN 40 B3
Jasper TN 53 E3
Jasper TX 63 D3
Jay OK 50 C1
Jeanerette LA 63 E4
Jean Lafitte LA 64 A4
Jefferson GA 53 F4
Jefferson IA 24 A4
Jefferson LA 111 B2
Jefferson OH 27 E4
Jefferson OR 17 E1
Jefferson SD 23 F3
Jefferson TX 62 C1
Jefferson WI 25 E3
Jefferson City MO 39 D3
Jefferson City TN 53 F2
Jeffersontown KY 40 C3
Jefferson Valley NY 70 A3
Jeffersonville IN 40 C3
Jeffersonville KY 41 D4
Jellico TN 53 F1
Jemez Pueblo NM 47 F2
Jemison AL 64 C1
Jena LA 63 E2
Jenison MI 26 A3
Jenkins KY 41 E4
Jenkintown PA 69 D2
Jenks OK 50 C2
Jennings LA 63 E3
Jennings MO 120 C1
Jensen Beach FL 67 F2
Jerome ID 19 E3
Jersey City NJ 69 E1
Jersey Shore PA 28 A4
Jerseyville IL 39 E3
Jesup GA 66 A2
Jesup IA 24 C3
Jewett City (Griswold) CT 71 D3
Jim Thorpe PA 68 C1
Joanna SC 54 B3
John Day OR 18 B1
Johnson AR 51 D2
Johnson VT 29 E1
Johnsonburg PA 27 F4
Johnson City KS 36 C4
Johnson City NY 28 B4
Johnson City TN 54 A1
Johnston SC 54 B4
Johnston City IL 39 F4
Johnstown CO 35 F1
Johnstown NY 28 C3
Johnstown OH 41 E2
Johnstown PA 42 B1
Joliet IL 25 F4
Joliet MT 11 E4
Jollyville TX 59 F3
Jones OK 50 B2
Jonesboro AR 51 F2
Jonesboro GA 53 F4
Jonesboro LA 63 E1
Jonesborough TN 54 A1
Jones Creek TX 61 E1
Jonestown MS 52 A3
Jonestown TX 59 F3
Jonesville LA 63 E2
Jonesville MI 26 B4
Joplin MO 39 D4
Joppatowne MD 68 B3
Jordan MN 24 B1
Joseph City AZ 46 C3

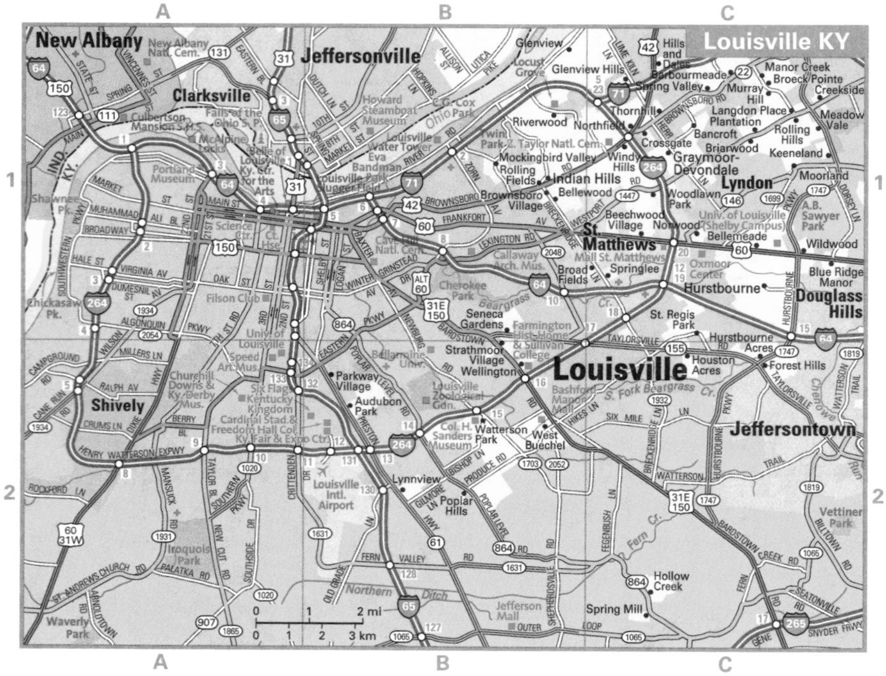

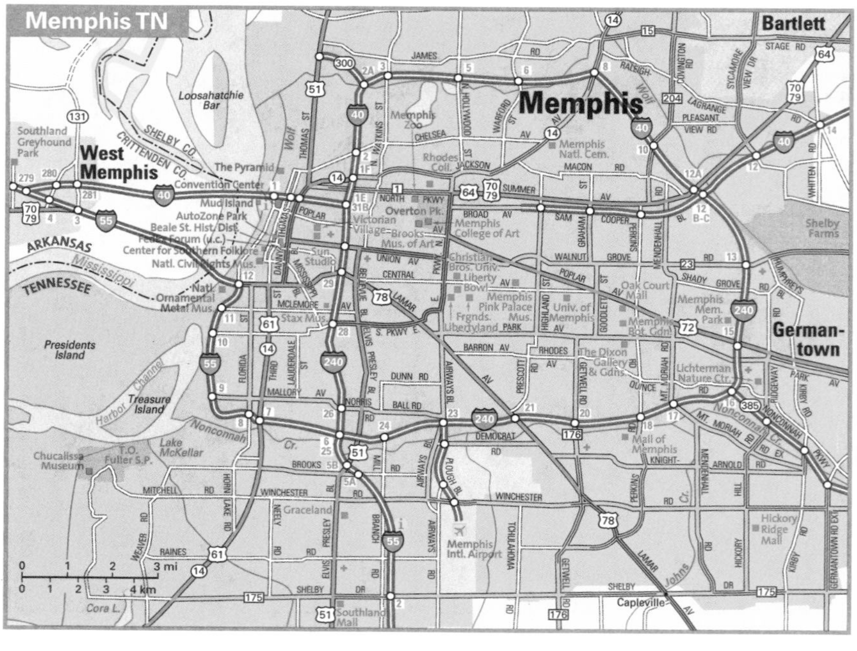

Miami/Fort Lauderdale FL

Milwaukee WI

Kiel WI	25 E2	Knoxville IA	24 B4
Kihei HI	73 D1	Knoxville TN	53 F2
Kilauea HI	72 B1	Kodiak AK	74 C4
Kilgore TX	62 C1	Kokomo IN	40 B1
Killdeer ND	12 C2	Konawa OK	50 B3
Kill Devil Hills NC	55 F2	Kooskia ID	9 F3
Killeen TX	59 F3	Kosciusko MS	64 A1
Kiln MS	64 A3	Kotlik AK	74 B2
Kimball NE	22 A4	Kotzebue AK	74 B1
Kimball SD	23 D2	Kountze TX	62 C3
Kimberling City MO	51 D1	Krebs OK	50 C3
Kimberly AL	53 D4	Kremmling CO	35 E2
Kimberly ID	19 E3	Kualapuu HI	72 B3
Kinder LA	63 D3	Kulm ND	13 E3
Kindred ND	13 F3	Kulpmont PA	68 B1
King NC	54 C1	Kuna ID	19 D2
King City CA	44 A1	Kutztown PA	68 C1
King Cove AK	74 A4	Kwethluk AK	74 B3
Kingfisher OK	50 A2	Kyle SD	22 B2
Kingman AZ	45 F2	Kyle TX	59 F4
Kingman KS	37 E4		
King of Prussia PA	68 C2	**L**	
Kings Beach CA	32 A2	Labadieville LA	63 F4
Kingsburg CA	44 B1	La Barge WY	20 B3
Kingsford MI	15 F4	La Belle FL	67 E3
Kingsland GA	66 B3	La Canada Flintridge	
Kingsland TX	59 F3	CA	106 D1
Kings Mtn. NC	54 B2	Lac du Flambeau WI	15 E4
Kingsport TN	54 A1	La Center WA	8 B4
Kingston ID	9 F2	La Cienega NM	47 F2
Kingston MA	71 E2	Lacey WA	8 B3
Kingston NH	71 E1	Lacombe LA	64 A3
Kingston NY	70 A2	La Crescent MN	24 C2
Kingston OK	50 B4	La Crescenta CA	106 C1
Kingston PA	28 B4	La Crosse KS	37 E3
Kingston RI	71 D3	La Crosse WI	24 C2
Kingston TN	53 E2	La Cygne KS	38 B3
Kingston Sprs. TN	52 C2	Ladonia AL	65 E1
Kingstree SC	54 C4	Ladson SC	66 C1
Kingsville TX	61 E3	Ladue MO	120 B2
Kingwood WV	42 A2	Lady Lake FL	67 E1
Kinnelon NJ	70 A4	Ladysmith WI	15 D4
Kinsey AL	65 E1	Lafayette AL	65 D1
Kinsley KS	37 D4	Lafayette CA	122 C1
Kinston NC	55 E2	Lafayette CO	35 F2
Kiowa KS	49 F1	La Fayette GA	53 E3
Kipnuk AK	74 A3	Lafayette IN	40 B1
Kirby TX	59 F4	Lafayette LA	63 E3
Kirbyville TX	63 D3	Lafayette OR	8 A4
Kirkland WA	8 B2	Lafayette TN	53 D1
Kirksville MO	38 C1	La Feria TX	61 D4
Kirkwood DE	68 C3	Lafitte LA	64 A4
Kirkwood MO	120 A3	La Follette TN	53 F1
Kirtland NM	47 E1	Lago Vista TX	59 F3
Kirtland OH	27 D4	La Grande OR	9 E4
Kiryas Joel NY	70 A3	LaGrange GA	65 E1
Kissimmee FL	67 E1	La Grange IL	98 C4
Kittanning PA	42 A1	Lagrange IN	26 A4
Kittery ME	29 F3	La Grange KY	40 C3
Kitty Hawk NC	55 F2	La Grange NC	55 E2
Klamath Falls OR	17 F3	La Grange TX	62 A4
Klawock AK	75 E4	Laguna Beach CA	44 C4
Knightstown IN	40 C2	Laguna Hills CA	107 G5
Knob Noster MO	38 C3		
Knox IN	26 A4		
Knoxville IL	39 E1		

Laguna Niguel CA	107 G6	Lakewood CO	35 F2
Laguna Woods CA	107 G5	Lakewood NJ	69 E2
La Habra CA	106 D3	Lakewood NY	42 A1
Lahaina HI	73 D1	Lakewood OH	99 E2
La Joya TX	60 C4	Lakewood TN	53 D1
La Junta CO	36 A4	Lakewood WA	8 F2
Lake Andes SD	23 D3	Lakewood Park FL	67 F3
Lake Arrowhead CA	45 D3	Lakewood Park ND	13 E2
Lake Arthur LA	63 E4	Lake Worth FL	67 F3
Lake Barcroft VA	124 B3	Lake Wylie SC	54 B3
Lake Carmel NY	70 B2	Lake Zurich IL	98 B1
Lake Charles LA	63 D3	Lakin KS	36 C4
Lake City AR	52 A2	Lakota ND	13 E1
Lake City FL	66 A3	La Luz NM	57 F1
Lake City IA	24 A3	Lamar AR	51 D2
Lake City MI	26 A1	Lamar CO	36 B4
Lake City MN	24 C1	Lamar MO	38 B4
Lake City PA	27 E4	La Marque TX	61 E1
Lake City SC	54 C4	Lambert MS	52 A3
Lake City TN	53 F1	Lambertville MI	26 C4
Lake Crystal MN	24 A2	Lambertville NJ	68 C2
Lake Dallas TX	62 A1	Lame Deer MT	11 F4
Lake Delton WI	25 D2	La Mesa CA	56 B4
Lake Elsinore CA	45 D4	La Mesa NM	57 F2
Lakefield MN	23 F2	Lamesa TX	58 C1
Lake Forest CA	107 G5	La Mirada CA	106 D3
Lake Forest IL	98 C1	Lamoni IA	38 C1
Lake Forest Park WA	123 B2	Lamont CA	44 C2
Lake Geneva WI	25 D3	La Moure ND	13 F3
Lake Grove NY	69 F1	Lampasas TX	59 F3
Lake Hamilton AR	51 E3	Lanai City HI	72 C4
Lake Havasu City AZ	45 F3	Lancaster CA	44 C3
Lakehills TX	59 E4	Lancaster KY	40 C4
Lake Isabella CA	44 C2	Lancaster NH	29 E1
Lake Jackson TX	61 E1	Lancaster OH	41 E2
Lake Junaluska NC	54 A2	Lancaster PA	68 B2
Lakeland FL	67 E2	Lancaster SC	54 B3
Lakeland GA	66 A2	Lancaster WI	25 D3
Lakeland TN	52 A2	Lancaster WY	20 C3
Lake Mary FL	67 E1	Land O' Lakes FL	67 D1
Lake Mills IA	24 B2	Landover MD	124 D1
Lake Mills WI	25 D3	Landrum SC	54 A3
Lake Montezuma AZ	46 B3	Lanett AL	65 E1
Lake Odessa MI	26 A3	Langdon ND	13 F1
Lake Orion MI	26 C3	Langley Park MD	124 D1
Lake Oswego OR	8 A3	Langston OK	50 B2
Lake Ozark MO	38 C3	Lanham MD	124 E1
Lake Panasoffkee FL	67 D1	Lanoka Harbor NJ	69 E2
Lake Placid NY	29 D1	Lansdale PA	68 C2
Lakeport CA	31 E3	Lansdowne MD	95 B3
Lake Preston SD	23 E1	Lansdowne PA	116 B3
Lake Providence LA	63 F1	L'Anse MI	15 F3
Lake St. Louis MO	39 E3	Lansford PA	68 C1
Lake Shore MD	56 B4	Lansing IL	98 E6
Lakeside CA	56 B4	Lansing KS	38 B2
Lakeside OR	17 D2	Lansing MI	26 B3
Lakeside VA	119 A1	Lansing NY	28 B2
Lakesite TN	53 E3	Lantana FL	67 F3
Lake Stevens WA	8 A3	La Palma CA	106 E4
Lakeview OR	18 A3	Lapeer MI	26 C3
Lake Vil. AR	51 F4	La Pine OR	17 F2
Lakeville MN	24 B1	Laplace LA	63 F4
Lake Wales FL	67 E2	La Plata MD	42 C3
Lakeway TX	59 F3	La Plata MO	39 D2
Lakewood CA	106 D4	Laporte CO	35 F1

Joshua TX	62 A1	Kalaoa HI	73 E3
Joshua Tree CA	45 E4	Kalispell MT	10 A1
Jourdanton TX	60 C1	Kalkaska MI	26 B1
Judsonia AR	51 F2	Kalona IA	24 C4
Julesburg CO	36 B3	Kamas UT	34 B1
Juliaetta ID	9 F3	Kamiah ID	9 F3
Junction TX	59 E3	Kanab UT	46 B1
Junction City KS	37 F3	Kane PA	27 F4
Junction City KY	40 C4	Kaneohe HI	72 A3
Junction City OR	17 E1	Kankakee IL	25 F4
Juneau AK	75 E4	Kannapolis NC	54 C2
Juneau WI	25 E2	Kansas City KS	38 B3
Juno Beach FL	67 F3	Kansas City MO	38 B2
Jupiter FL	67 F3	Kapaa HI	72 B1
Justin TX	62 A1	Kaplan LA	63 E4
		Karnes City TX	61 D1
K		Kasigluk AK	74 A3
Kadoka SD	22 B2	Kasson MN	24 B2
Kahaluu HI	72 A3	Kathleen FL	67 E2
Kahoka MO	39 D1	Katy TX	61 E1
Kahuku HI	72 A2	Kaufman TX	62 B1
Kahului HI	73 D1	Kaukauna WI	25 E1
Kaibito AZ	46 C1	Kaunakakai HI	72 B3
Kailua HI	72 A3	Kayenta AZ	46 C1
Kailua-Kona HI	73 E3	Kaysville UT	34 A1
Kake AK	75 E4	Keaau HI	73 E3
Kalaheo HI	72 B1	Kealakekua HI	73 E3
Kalamazoo MI	26 A3	Kearney MO	38 B2
		Kearney NE	37 E1

Kearny AZ	56 B1	Kenova WV	41 E3
Kearny NJ	112 B3	Kensett AR	51 F2
Keauhou HI	73 E3	Kent OH	27 D4
Keene NH	70 C1	Kent WA	8 B2
Keene TX	62 A1	Kenton OH	41 D1
Keizer OR	17 E1	Kentwood LA	63 F3
Kekaha HI	72 A1	Kentwood MI	26 A3
Keller TX	100 D1	Kenyon MN	24 B1
Kellogg ID	9 F2	Keokuk IA	39 D1
Kelso WA	8 B3	Kerens TX	62 B2
Kemmerer WY	20 B4	Kerman CA	44 B1
Kenai AK	74 C3	Kermit TX	58 B2
Kendall FL	67 F4	Kernersville NC	54 C2
Kendallville IN	26 A4	Kerrville TX	59 E4
Kenedy TX	61 D2	Kersey CO	36 A1
Kenesaw NE	37 E1	Kershaw SC	54 B3
Kenmare ND	12 C1	Ketchikan AK	75 E4
Kenmore NY	97 B2	Ketchum ID	19 E2
Kenmore WA	123 B2	Kettering OH	41 D2
Kennebunk ME	29 F2	Kettle Falls WA	9 E1
Kennebunkport ME	29 F2	Kewanee IL	25 D4
Kenner LA	64 A4	Kewaskum WI	25 E2
Kennesaw GA	53 E4	Kewaunee WI	25 F1
Kennett MO	52 A1	Key Biscayne FL	67 F4
Kennett Square PA	68 C3	Key Largo FL	67 E4
Kennewick WA	9 D3	Keyes CA	31 F4
Kenosha WI	25 F3	Keyport NJ	69 E1
		Keyser WV	42 B2
		Key West FL	67 E4

La Porte IN 25 F4	Lavaca AR 51 D2	Leeds AL 53 D4	Levittown PA 69 D2	Lidgerwood ND 13 F3	Linton IN 40 B3	Livonia MI 26 C3	Long Beach CA 44 B4
La Porte TX 61 F1	Lava Hot Sprs. ID 20 A3	Leeds ND 13 E1	Lewes DE 69 D4	Ligonier IN 26 A4	Linton ND 13 D3	Llano TX 59 F3	Long Beach MS 64 A4
La Porte City IA 24 C3	La Vergne TN 53 D2	Leesburg FL 67 E1	Lewisburg PA 42 C1	Lihue HI 72 B1	Linwood NJ 69 D3	Lochearn MD 95 A2	Long Beach NY 69 F1
La Pryor TX 60 B1	La Verkin UT 33 F4	Leesburg GA 65 E2	Lewisburg TN 53 D2	Lilburn GA 53 F4	Lisbon IA 24 C4	Lockeford CA 31 F3	Longboat Key FL 67 D2
La Puente CA 106 E3	La Verne CA 107 F3	Leesburg VA 68 A4	Lewisburg WV 41 F4	Lillington NC 55 D2	Lisbon ME 29 F2	Lockhart TX 62 A4	Longbranch MA 70 C2
Lapwai ID 9 E3	Lawai HI 72 B1	Lees Summit MO 38 B3	Lewisport KY 40 B4	Lima OH 41 D1	Lisbon ND 13 F3	Lock Haven PA 42 B1	Long Branch NJ 69 E2
Laramie WY 21 E4	Lawndale CA 106 C3	Leesville LA 63 D3	Lewiston ID 9 E3	Limestone ME 30 C1	Lisbon OH 41 F1	Lockney TX 49 D4	Longmeadow MA 71 E1
Laredo TX 60 B3	Lawrence IN 40 B2	Lehigh Acres FL 67 E3	Lewiston ME 29 F2	Limon CO 36 A2	Lisbon Falls ME 29 F2	Lockport IL 98 B6	Longmont CO 35 F1
Largo FL 67 D2	Lawrence KS 38 B3	Lehighton PA 68 C1	Lewiston MN 24 C2	Lincoln AL 53 D4	Lisle IL 98 B5	Lockport LA 63 F4	Long Prairie MN 14 B4
Larimore ND 13 F2	Lawrence MA 71 E1	Leitchfield KY 40 B4	Lewiston UT 20 A4	Lincoln AR 51 D2	Litchfield IL 39 F2	Lockport NY 27 E3	Longview TX 62 C1
Larned KS 37 E3	Lawrenceburg IN 40 C3	Leland MS 51 F4	Lewistown IL 39 E1	Lincoln CA 31 F2	Litchfield MN 14 B4	Locust NC 54 C2	Longview WA 8 B3
Larose LA 63 F4	Lawrenceburg KY 40 C4	Lemay MO 120 B3	Lewistown MT 11 D3	Lincoln IL 39 F1	Lithia Sprs. GA 53 E4	Locust Grove GA 53 F4	Lonoke AR 51 F3
La Salle CO 35 F1	Lawrenceburg TN 52 C2	Le Mars IA 23 F3	Lewistown PA 68 A1	Lincoln KS 37 E3	Lithonia GA 53 F4	Locust Grove OK 50 C2	Lonsdale MN 24 B1
La Salle IL 25 E4	Lawrenceville GA 53 F4	Lemmon SD 12 C4	Lexington KY 40 C4	Lincoln ME 30 B3	Lititz PA 68 B2	Lodge Grass MT 11 F4	Loogootee IN 40 B3
Las Animas CO 36 B4	Lawrenceville IL 40 A3	Lemmon Valley NV 32 B2	Lexington MA 71 E1	Lincoln MT 10 B3	Lodi CA 31 F3	Lodi CA 31 F3	Loomis CA 31 F3
Las Cruces NM 57 E1	Lawrenceville NJ 69 D2	Lemon Grove CA 56 B4	Lexington MO 38 C2	Lincoln ND 13 D3	Lodi OH 41 E1	Logan IA 23 F4	Lorain OH 27 D4
Las Vegas NV 45 E2	Lawson MO 38 B2	Lemoore CA 44 C2	Lexington NE 37 D1	Lincoln NE 37 F1	Lodi CA 112 C1	Logan NM 48 C2	Lordsburg NM 57 D1
Las Vegas NM 48 A3	Lawton OK 49 F3	Lena IL 25 D3	Lexington NC 54 C2	Lincoln Beach OR 17 D1	Lodi OH 41 E1	Logan OH 41 E2	Lorena TX 62 A2
Lathrop CA 31 F4	Layton UT 34 A1	Lenexa KS 104 A4	Lexington OH 41 E1	Lincoln City OR 17 D1	Lodi WV 41 F1	Logan UT 20 A4	Loretto TN 52 C3
Lathrop MO 38 B2	Lead SD 21 F1	Lennox CA 106 C3	Lexington OK 50 A3	Lincoln ME 30 B3	Lofall WV 41 F1	Logan WV 41 F4	Loris SC 55 D4
Latimer MS 64 B3	Leadville CO 35 E2	Lennox SD 23 E2	Lexington SC 54 B4	Lincolnia VA 124 B3	Little Canada MN 110 D2	Logandale NV 45 F1	Los Alamos NM 47 F2
Latrobe PA 42 A1	League City TX 61 F1	Lenoir NC 54 B2	Lexington TN 52 C2	Lincolnton NC 54 B2	Little Chute WI 25 E1	Logansport IN 40 B1	Los Banos CA 44 A1
Lauderdale Lakes FL 109 B2	Lealman FL 123 D2	Lenoir City TN 53 F2	Lexington VA 42 A4	Lincoln Vil. OH. 99 A3	Little Cypress TX 63 D3	Logansport LA 63 D2	Los Chavez NM 47 F4
Lauderhill FL 109 A2	Leander TX 59 F3	Lenox IA 38 B1	Lexington SC 54 B4	Linda CA 31 F2	Little Falls MN 14 B4	Loganville GA 53 F4	Los Fresnos TX 61 D4
Laughlin NV 45 F2	Leavenworth KS 38 B2	Lenox MA 71 D1	Lexington TN 52 C2	Lindale GA 53 E4	Little Falls NJ 112 A1	Logan NV 45 F1	Los Gatos CA 31 F4
La Union NM 57 E2	Leavenworth WA 8 C2	Lenwood CA 45 D3	Lexington VA 42 A4	Lindale TX 62 B1	Little Falls NY 28 C3	Lolo MT 10 A3	Los Lunas NM 47 F3
Laurel DE 43 D3	Lebanon IN 40 B2	Leo-Cedarville IN 26 B4	Lexington Park MD 42 C3	Linden AL 64 C1	Little Ferry NJ 112 C2	Loma Linda CA 107 K3	Los Osos CA 44 A2
Laurel FL 67 D2	Lebanon KY 40 C4	Leominster MA 71 D1	Libby MT 9 F1	Linden MI 26 C3	Littlefield TX 48 C4	Lombard IL 98 B4	Loudon TN 53 F2
Laurel MD 68 A4	Lebanon MO 38 C4	Leon IA 38 C1	Liberal KS 49 D1	Linden NJ 112 A5	Littleton CO 35 F2	Lomira WI 25 E2	Loudonville NY 70 B1
Laurel MS 64 B2	Lebanon NH 29 E2	Leon Valley TX 59 F4	Liberty IN 40 C2	Linden TX 62 C1	Littleton NH 29 E1	Lomita CA 106 C4	Loudonville OH 41 E1
Laurel MT 11 E4	Lebanon OH 41 D2	Leonard TX 50 B4	Liberty KY 40 C4	Lindenhurst NY 69 F1	Live Oak CA 31 F2	Lompoc CA 44 A3	Louisa KY 41 E4
Laurel NE 23 E3	Lebanon OR 17 E1	Leonia NJ 112 D1	Liberty MO 38 B2	Lindenwold NJ 69 D3	Live Oak FL 65 F3	London KY 53 F1	Louisa VA 42 B4
Laurel VA 42 C4	Lebanon PA 68 B2	Le Roy IL 39 F1	Liberty NY 28 C4	Lindsay CA 44 C2	Live Oak KY 40 C4	London OH 41 D2	Louisburg KS 38 B3
Laurel Bay SC 66 B1	Lebanon TN 53 D2	Le Roy NY 27 E3	Liberty SC 54 A3	Lindsay OK 50 A3	Livermore CA 31 F4	Londonderry NH 71 D1	Louisburg NC 55 D2
Laureldale PA 68 C2	Lebanon VA 54 A1	Leslie MI 26 B3	Liberty TX 62 C4	Lindsborg KS 37 F3	Livermore KY 40 B4	Londonium MO 39 E2	Louisiana MO 39 E2
Laurens IA 24 A3	Lebanon Jct. KY 40 C4	Le Sueur MN 24 B1	Liberty Hill TX 59 F3	Lindstrom MN 14 C4	Livermore Falls ME 29 F1	Londonum MD 68 A3	Louisville CO 101 A3
Laurens SC 54 A3	Lecanto FL 67 D1	Levelland TX 48 C4	Libertyville IL 98 C1	Lineville AL 53 E4	Livingston AL 64 B1	Lone Grove OK 50 B4	Louisiville GA 66 A1
Laurinburg NC 54 C3	Le Center MN 24 B1	Levittown NY 69 F1	Licking MO 39 D4	Lino Lakes MN 110 D1	Livingston CA 32 A4	Lone Star TX 62 C1	Louisville KY 40 B3
Laurium MI 15 F2	Le Claire IA 25 D4				Livingston MT 11 D4		Louisville MS 64 B1

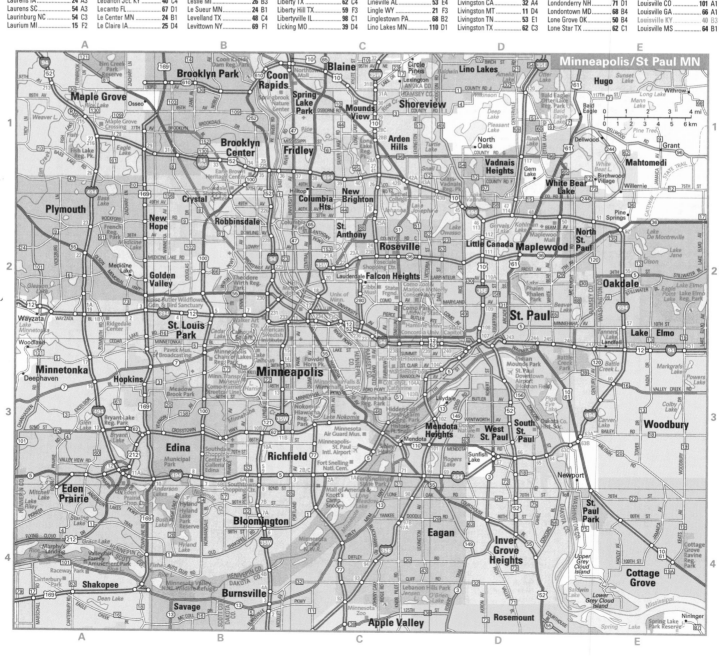

Minneapolis/St Paul MN

Entries in **bold color** indicate cities with detailed inset maps.
Louisville – Minot 111

Louisville NE 38 A1
Louisville OH 41 F1
Loup City NE 23 D4
Loveland CO 35 F1
Loveland OH 41 D2
Lovell WY 20 C1
Lovelock NV 32 B1
Loves Park IL 25 E3
Loving NM 58 A1
Lovington NM 58 B1
Lowell AR 51 D1
Lowell IN 25 F4
Lowell MA 71 E1
Lowell MI 26 A3
Lower Brule SD 23 D2
Lowville NY 28 C2
Lubbock TX 49 D4
Lucedale MS 64 B3
Lucerne WY 21 D2
Ludington MI 26 A2
Ludlow MA 70 C2
Ludlow VT 29 E2
Lufkin TX 62 C2
Lugoff SC 54 B3
Lukachukai AZ 47 D1
Luling LA 111 A2
Luling TX 61 D1
Lumberton MS 64 A3
Lumberton NC 55 D3
Lumberton TX 62 C3
Luray VA 42 B3
Lusk WY 21 F3
Lutcher LA 63 F4
Lutz FL 67 D1
Luverne AL 65 D2
Luverne MN 23 F2
Luxemburg WI 25 F1
Luxora AR 52 A2
Lyford TX 61 D4
Lyman WY 20 B4
Lynchburg TN 53 D2
Lynchburg VA 42 A4
Lynden WA 8 B1
Lyndhurst NJ 112 B2
Lyndhurst OH 99 G2
Lyndhurst VA 42 B4
Lyndon KS 38 A3
Lyndon KY 108 C1
Lyndonville VT 29 E1
Lynn MA 71 E1
Lynn Haven FL 65 D3
Lynnwood WA 8 B2
Lynwood CA 106 D3
Lyons CO 35 F1
Lyons GA 66 A1
Lyons KS 37 E3
Lyons NE 23 F4
Lyons NY 28 B3
Lytle TX 59 F4

M

Mabank TX 62 B1
Mableton GA 53 E4
Mabton WA 8 C3
Macclenny FL 66 A3
Macedonia OH 99 G3
Machesney Park IL 25 E3
Machias ME 30 C3
Mackay ID 19 F2
Macomb IL 39 E1
Macon GA 65 F1
Macon MS 64 B1
Macon MO 39 D2
Macungie PA 68 C1
Macy NE 23 F3
Madawaska ME 30 B1
Maddock ND 13 E2
Madeira OH 99 C2
Madeira Beach FL 67 D2
Madelia MN 24 A2
Madera CA 44 B1
Madill OK 50 B4
Madison AL 53 D3
Madison FL 65 F3
Madison GA 53 F4
Madison IN 40 C3
Madison ME 29 F1
Madison MN 23 F1
Madison MS 64 A1
Madison NE 23 E4
Madison NJ 69 E1
Madison NC 54 C1
Madison SD 23 E2
Madison WV 41 E4
Madison WI 25 E3
Madison Hts. MI 101 G1
Madisonville KY 40 A4
Madisonville TN 53 F2
Madisonville TX 62 B3
Madras OR 17 F1
Madrid IA 24 B4
Maeser UT 34 C1
Magalia CA 31 F2
Magdalena NM 47 F4
Magee MS 64 A2
Magna UT 34 A1
Magnolia AR 51 E4
Magnolia MS 63 F3
Mahanoy City PA 68 B1
Mahomet IL 39 F1
Mahopac NY 70 B3
Mahwah NJ 70 A4
Maiden NC 54 B2
Maili HI 72 A3
Maize KS 37 F4
Makaha HI 72 A3
Makakilo City HI 72 A3
Makawao HI 73 D1
Makena HI 73 D1
Malad City ID 20 A3
Malakoff TX 62 B2
Malden MA 71 E1
Malden MO 52 A1
Malibu CA 44 C4
Malone NY 28 C1
Malta MT 11 E1
Malvern AR 51 E3
Malvern PA 68 C2
Mammoth AZ 56 B1
Mammoth Lakes CA 32 B4
Mamou LA 63 E3
Manasquan NJ 69 E2
Manassa CO 35 F4
Manassas VA 68 A4
Manassas Park VA 68 A4
Manchester CT 70 C3
Manchester GA 65 E1
Manchester IA 24 C3
Manchester KY 53 F1
Manchester MD 68 A3
Manchester MI 26 B3
Manchester MO 120 A3
Manchester NH 29 E3
Manchester TN 53 D2
Manchester-by-the-Sea MA 71 E1
Mancos CO 35 D4
Mandan ND 13 D3
Mandaree ND 12 C2
Manderson SD 22 B2
Mandeville LA 64 A3
Mangum OK 49 E3
Manhattan IL 25 F4
Manhattan KS 37 F2
Manhattan MT 10 C4
Manhattan Beach CA 106 B3
Manheim PA 68 B2
Manila AR 52 A2
Manistee MI 26 A1
Manistique MI 16 B4
Manitou Beach MI 26 B4
Manitou Sprs. CO 35 F3
Manitowoc WI 25 F2
Mankato MN 24 B1
Manlius NY 28 B3
Mannford OK 50 B2
Manning IA 24 A4
Manning SC 54 C4
Mannington WV 41 F2
Manomet MA 71 E2
Mansfield LA 63 D2
Mansfield MA 71 E2
Mansfield OH 41 E1
Mansfield PA 28 B4
Mansfield TX 62 A1
Manson IA 24 A3
Mansura LA 63 E3
Mantachie MS 52 B3
Manteca CA 31 F4
Manteno IL 25 F4
Manti UT 34 B2
Manvel TX 61 F1
Manville NJ 69 D1
Manville RI 71 D2
Many LA 63 D2
Many Farms AZ 47 D1
Maple Grove MN 14 C4
Maple Hts. OH 99 G3
Maple Lake MN 14 C4
Mapleton IA 23 F3
Mapleton MN 24 B2
Mapleton ND 13 F3
Mapleton UT 34 B2
Maple Valley WA 8 B2
Maplewood MN 110 D2
Maplewood MO 120 B2
Maquoketa IA 25 D4
Marana AZ 56 B1
Marathon FL 67 D4
Marathon City WI 25 D1
Marble Falls TX 59 F3
Marblehead MA 71 E1
Marble Hill MO 39 E4
Marbleton WY 20 B3
Marceline MO 38 C2
Marco Island FL 67 E4
Marengo IL 25 E3
Marengo IA 24 C4
Marfa TX 58 A3
Margate FL 109 A1
Margate City NJ 69 E3
Marianna AR 51 F3
Marianna FL 65 E3
Maricopa AZ 56 A1
Marietta GA 53 E4
Marietta OH 41 F2
Marietta OK 50 B4
Marietta PA 68 B2
Marietta SC 54 A3
Marietta WA 8 B1
Marina CA 31 D4
Marine City MI 26 C3
Marinette WI 16 A4
Marion AL 64 C1
Marion AR 52 A2
Marion IL 39 F4
Marion IN 40 C1
Marion IA 24 C4
Marion KS 37 F3
Marion KY 40 A4
Marion MS 64 B1
Marion NC 54 A2
Marion OH 41 D1
Marion SC 54 C4
Marion SD 23 E2
Marion VA 54 B1
Marionville MO 51 D1
Marked Tree AR 52 A2
Marks MS 52 A3
Marksville LA 63 E3
Marlborough MA 71 D2
Marlette MI 26 C2
Marlin TX 62 A2
Marlinton WV 42 A3
Marlow OK 50 A3
Marlton NJ 69 D2
Marmet WV 41 F3
Marquette MI 16 A3
Marrero LA 64 A4
Marseilles IL 25 E4
Marshall AR 51 E2
Marshall IL 40 A2
Marshall MI 26 B3
Marshall MN 23 F1
Marshall MO 38 C3
Marshall TX 62 C1
Marshalltown IA 24 B4
Marshfield MA 71 E2
Marshfield MO 38 C4
Marshfield WI 25 D1
Mars Hill ME 30 C1
Marshville NC 54 C3
Marsing ID 19 D2
Mart TX 62 A2
Martha Lake WA 123 B2
Martin SD 22 B2
Martin TN 52 B1
Martinez CA 31 F4
Martinsburg WV 42 B2
Martins Ferry OH 41 F2
Martinsville IN 40 B2
Martinsville VA 54 C1
Marvell AR 51 F3
Mary Esther FL 64 C3
Maryland Hts. MO 120 A1
Marysville CA 31 F2
Marysville KS 37 F2
Marysville MI 26 C3
Marysville OH 41 D1
Marysville WA 8 B2
Maryville MO 38 B1
Maryville TN 53 F2
Mascot TN 53 F2
Mascoutah IL 39 E3
Mason MI 26 B3
Mason NV 32 B3
Mason OH 41 D2
Mason TX 59 E3
Mason City IL 39 F1
Mason City IA 24 B2
Masontown PA 42 A2
Massapequa NY 69 F1
Massena NY 28 C1
Massillon OH 41 F1
Mastic NY 43 F1
Mastic Beach NY 43 F1
Matawan NJ 69 E2
Mathews LA 63 F4
Mathis TX 61 D2
Mattapoisett MA 71 E3
Mattawa WA 8 C3
Mattawan MI 26 A3
Matthews NC 54 B3
Mattituck NY 70 C4
Mattoon IL 39 F2
Maud OK 50 B3
Mauldin SC 54 A3
Maumee OH 26 C4
Maumelle AR 51 E3
Maunawili HI 72 A3
Mauriceville TX 63 D4
Mauston WI 25 D2
Maxton NC 54 C3
Maybrook NY 70 A3
Mayer AZ 46 B3
Mayfield KY 52 B1
Mayfield Hts. OH 99 G2
Mayflower AR 51 E3
Maynard MA 71 D1
Maynardville TN 53 F1
Mayo MD 68 B4
Mayo SC 54 B3
Mayodan NC 54 C1
Maysville KY 41 D3
Maysville OK 50 A3
Mayville ND 13 F2
Mayville WI 25 E2
Maywood CA 106 D3
Maywood IL 98 C4
Maywood NJ 112 C1
McAlester OK 50 C3
McAllen TX 60 C4
McCall ID 19 D1
McCamey TX 58 C3
McCammon ID 20 A3
McCandless PA 42 A1
McClusky ND 13 D2
McColl SC 54 C3
McComb MS 63 F2
McCook NE 37 D1
McCrory AR 51 F2
McDonough GA 53 F4
McEwen TN 52 C2
McFarland CA 44 B2
McFarland WI 25 E3
McGehee AR 51 F4
McGill NV 33 E2
McGregor TX 62 A2
McHenry IL 25 E3
McKeesport PA 42 A1
McKenzie TN 52 B2
McKinleyville CA 31 D1
McKinney TX 50 B4
McLaughlin SD 13 D4
McLean VA 68 A4
McLeansboro IL 39 F4
McLoud OK 50 B2
McMechen WV 41 F2
McMinnville OR 8 A4
McMinnville TN 53 D2
McPherson KS 37 F3
McRae GA 65 F2
McSherrystown PA 68 A3
McVille ND 13 F2
Mead CO 35 F1
Mead WA 9 E2
Meade KS 37 D4
Meadowview VA 54 B1
Meadville PA 27 E4
Mebane NC 55 D2
Mecca CA 45 E4
Mechanic Falls ME 29 F1
Mechanicsburg PA 68 A2
Mechanicsville VA 42 C4
Mechanicville NY 70 B1
Medfield MA 71 E2
Medford MA 29 F3
Medford NJ 69 D2
Medford NY 69 F1
Medford OK 50 A1
Medford OR 17 E3
Medford WI 15 E4
Medford Lakes NJ 69 D3
Media PA 68 C2
Mediapolis IA 39 D1
Medical Lake WA 9 E2
Medicine Lodge KS 37 E4
Medina NY 27 F2
Medina OH 41 E1
Medulla FL 67 E2
Meeker CO 35 D2
Mehlville MO 120 B3
Meiners Oaks CA 44 B3
Melbourne AR 51 F2
Melbourne FL 67 F2
Melrose MN 14 B4
Melrose NM 48 B3
Melrose Park IL 98 C4
Melvindale MI 101 F3
Memphis MO 39 D1
Memphis TN 52 A3
Memphis TX 49 E3
Mena AR 51 D3
Menan ID 20 A2
Menard TX 59 E3
Menasha WI 25 E1
Mendenhall MS 64 A2
Mendham NJ 69 D1
Mendota CA 44 A1
Mendota IL 25 E4
Mendota Hts. MN 110 D3
Menlo Park CA 122 C5
Menno SD 23 E2
Menominee MI 16 A4
Menomonee Falls WI 25 E2
Menomonie WI 24 C1
Mentone CA 45 D4
Mentor OH 27 D4
Mequon WI 25 F2
Meraux LA 111 D2
Merced CA 32 A4
Mercedes TX 60 C4
Mercer Island WA 123 B3
Mercerville NJ 69 D2
Meredith NH 29 E2
Meriden CT 70 C3
Meridian ID 19 D2
Meridian MS 64 B1
Meridian PA 42 A1
Meridian TX 59 F2
Meridianville AL 53 D3
Merkel TX 59 E1
Merriam KS 104 B4
Merrifield VA 124 A2
Merrill WI 15 E4
Merrillville IN 25 F4
Merrimack NH 71 D1
Merritt Island FL 67 F1
Mesa AZ 46 B4
Mescalero NM 48 A4
Mesilla NM 57 E1
Mesita NM 47 E3
Mesquite NV 45 F1
Mesquite NM 57 E2
Mesquite TX 62 B1
Metairie LA 64 A4
Metamora IL 39 F1
Metcalfe MS 51 F4
Methuen MA 71 E1
Metlakatla AK 75 E4
Metropolis IL 39 F4
Metter GA 66 A1
Mexia TX 62 B2
Mexico ME 29 F1
Mexico MO 39 D3
Miami AZ 46 C4
Miami FL 67 F4
Miami OK 50 C1
Miami Beach FL 67 F4
Miami Lakes FL 109 A3
Micco FL 67 F2
Michigan Ctr. MI 26 B3
Michigan City IN 25 F4
Middleboro MA 71 E2
Middleburg FL 66 A4
Middleburg Hts. OH 99 E3
Middlebury IN 26 A4
Middlebury VT 29 D2
Middle River MD 68 B3
Middlesboro KY 53 F1
Middlesex NJ 69 E1
Middleton ID 19 D2
Middleton WI 25 E3
Middletown CT 70 C3
Middletown DE 68 C3
Middletown IN 40 C2
Middletown KY 40 C3
Middletown MD 68 A3
Middletown NY 70 A3
Middletown OH 40 C2
Middletown PA 68 B2
Middletown RI 71 E3
Middleville MI 26 A3
Midland MI 26 B2
Midland PA 41 F1
Midland TX 58 C2
Midland City AL 65 D2
Midlothian TX 62 A1
Midvale UT 120 E3
Midway KY 40 C4
Midway NM 48 B4
Midwest WY 21 E2
Midwest City OK 50 A2
Mifflinburg PA 68 A1
Milaca MN 14 C4
Milan MI 26 C4
Milan MO 38 C1
Milan NM 47 E3
Milan TN 52 B2
Milbank SD 14 A4
Miles City MT 12 A3
Milford CT 70 B4
Milford DE 68 C4
Milford IA 24 A2
Milford ME 30 B3
Milford MD 95 A2
Milford MA 71 D2
Milford MI 26 C3
Milford NE 37 F1
Milford NH 71 D1
Milford UT 33 F3
Mililani Town HI 72 A3
Millbrae CA 122 B4
Millbrook AL 65 D1
Millbury MA 71 D2
Mill City OR 17 E1
Mill Creek WA 123 B2
Milledgeville GA 65 F1
Millen GA 66 A1
Miller SD 23 D1
Millersburg OH 41 E1
Millersburg PA 68 A1
Millersville PA 68 B2
Millersville TN 53 D1
Milliken CO 35 F1
Millington TN 52 A2
Millinocket ME 30 B2
Mills WY 21 E3
Millsboro DE 43 D3
Mill Valley CA 31 E4
Millville NJ 69 D3
Milnor ND 13 F3
Milo ME 30 B3
Milpitas CA 31 F4
Milton DE 68 C4
Milton FL 64 C3
Milton MA 96 E3
Milton PA 42 C1
Milton VT 29 D1
Milton WV 41 E3
Milton WI 25 E3
Milton-Freewater OR 9 D4
Milwaukee WI 25 F3
Milwaukie OR 118 B2
Mims FL 67 F1
Minatare NE 22 A4
Minco OK 50 A2
Minden LA 63 D1
Minden NE 37 E1
Minden NV 32 B3
Mineola NY 69 F1
Mineola TX 62 B1
Mineral Pt. WI 25 D3
Mineral Sprs. AR 51 D4
Mineral Wells TX 59 F1
Mineral Wells WV 41 F2
Minersville PA 68 B1
Minerva OH 41 F1
Minneapolis KS 37 F3
Minneapolis MN 24 B1
Minneota MN 23 F1
Minnetonka MN 110 A3
Minooka IL 25 E4
Minot ND 13 D1

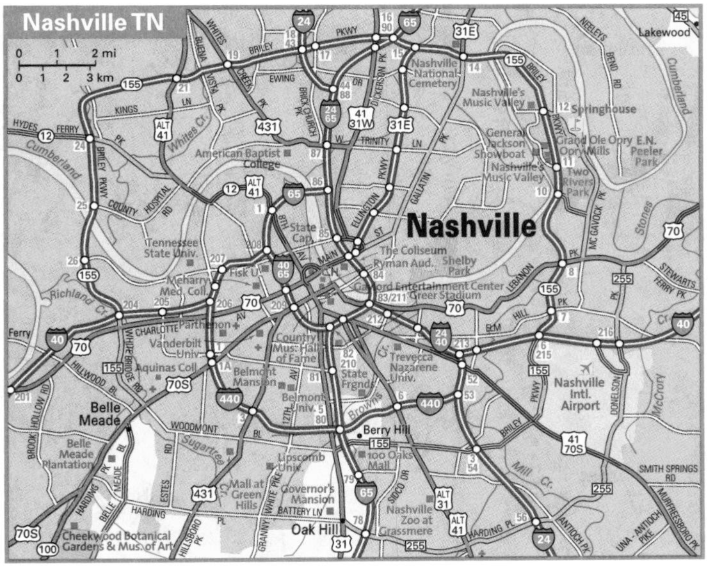

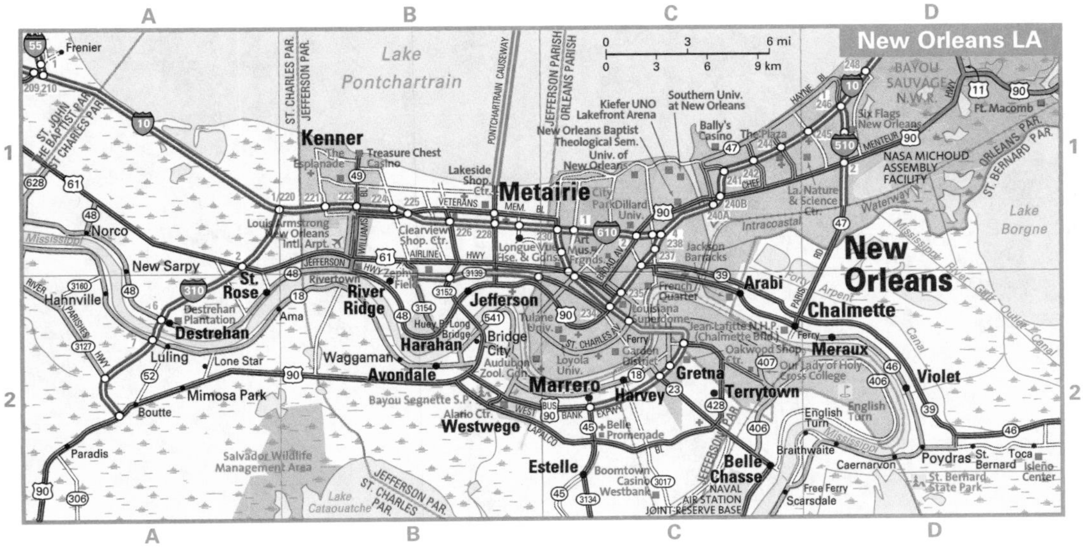

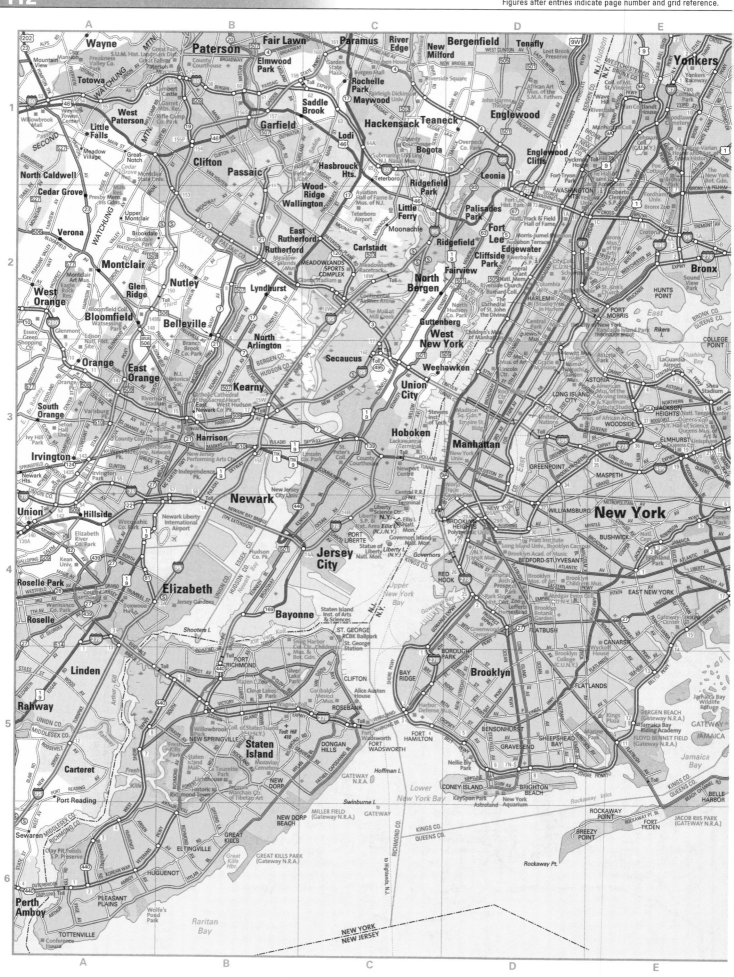

Entries in bold color indicate cities with detailed inset maps.

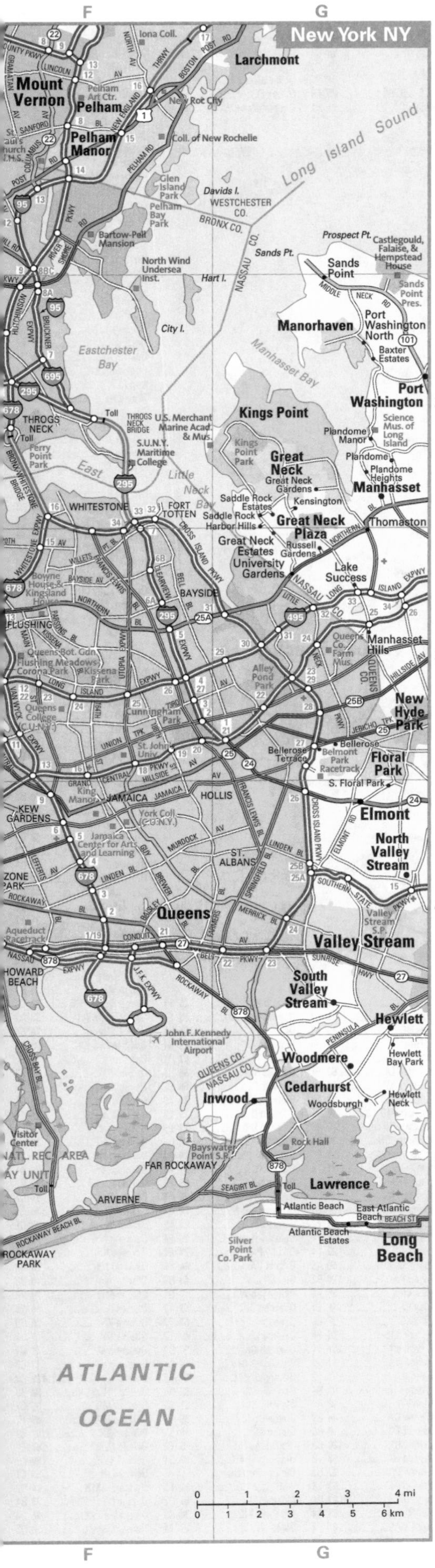

Minster OH 41 D1
Mint Hill NC 54 B3
Minto ND 13 F1
Minturn CO 35 E2
Mio MI 26 B1
Mira Loma CA 107 H3
Miramar FL 67 F4
Mishawaka IN 26 A4
Mission KS 104 B3
Mission SD 22 C2
Mission TX 60 C4
Mission Bend TX 102 C2
Mission Viejo CA 44 C4
Mississippi State MS 52 B4
Missoula MT 10 A3
Missouri City TX 61 E1
Missouri Valley IA 23 F4
Mitchell IN 40 B3
Mitchell NE 22 A4
Mitchell SD 23 E2
Moab UT 34 C3
Moapa NV 45 F1
Moberly MO 39 D2
Mobile AL 64 B3
Mobridge SD 13 D4
Mocksville NC 54 C2
Modesto CA 31 F4
Mohall ND 13 D1
Mohave Valley AZ 45 F3
Mohawk NY 28 C3
Mojave CA 44 C3
Mokuleia HI 72 A2
Molalla OR 8 B4
Moline IL 25 D4
Momence IL 25 F4
Monahans TX 58 B2
Moncks Corner SC 66 C1
Mondovi WI 24 C1
Monee IL 25 F4
Monessen PA 42 A1
Monett MO 51 D1
Monmouth IL 39 E1
Monmouth OR 17 E1
Monmouth Beach NJ 69 E2
Monona IA 24 C3
Monona WI 25 E3
Monroe CT 70 B3
Monroe GA 53 F4
Monroe IA 24 B4
Monroe LA 63 E1
Monroe MI 26 C4
Monroe NY 70 A3
Monroe NC 54 C3
Monroe OH 41 D2
Monroe UT 34 A3
Monroe WA 8 B2
Monroe WI 25 D3
Monroe City MO 39 D2
Monroeville AL 64 C2
Monroeville PA 42 A1
Monrovia CA 106 E2
Montague MI 26 A2
Montana City MT 10 C3
Montauk NY 71 D4
Mont Belvieu TX 62 C4
Montclair CA 107 G2
Montclair NJ 112 A2
Monte Alto TX 60 C4
Montebello CA 106 D3
Montecito CA 44 B3
Montegut LA 63 F4
Monterey CA 31 D4
Monterey TN 53 E2
Monterey Park CA 106 D2
Montesano WA 8 A3
Montevallo AL 64 C1
Montevideo MN 23 F1
Monte Vista CO 35 F4
Montezuma GA 65 F1
Montezuma IA 24 B4
Montgomery AL 65 D1
Montgomery MN 24 B1
Montgomery NY 70 A3
Montgomery OH 99 C1
Montgomery WV 41 F3
Montgomery City MO 39 D3
Montgomery Vil. MD 68 A4
Montgomeryville PA 69 D2
Monticello AR 51 F4
Monticello GA 53 F4
Monticello IL 39 F2
Monticello IN 40 B1
Monticello IA 24 C3
Monticello KY 53 E1
Monticello MN 14 C4
Monticello MS 64 A2
Monticello NY 28 C4
Monticello UT 34 C4
Montoursville PA 28 B4
Montpelier ID 20 A3
Montpelier OH 26 B4
Montpelier VT 29 E1
Montrose CA 106 D1
Montrose CO 35 D3
Monument CO 35 F2
Moody TX 62 A2
Moorcroft WY 21 F1
Moore OK 50 A2
Moorefield WV 42 B3
Mooreland OK 49 F1
Moorestown NJ 43 D2
Mooresville IN 40 B2
Mooresville NC 54 B2
Moorhead MN 14 A3
Moorhead MS 52 A4
Moorpark CA 44 C3
Moose Lake MN 14 C3
Moosup CT 71 D3
Mora MN 14 C4
Moraga CA 122 D2
Morehead KY 41 D3
Morehead City NC 55 E3
Morenci AZ 56 C1
Morenci MI 26 B4
Moreno Valley CA 45 D4
Morgan UT 34 B1
Morgan City LA 63 F4
Morganfield KY 40 A4
Morgan Hill CA 31 F4
Morganton NC 54 B2
Morgantown KY 40 B4
Morgantown WV 42 A2
Moriarty NM 47 F3
Moroni UT 34 B2
Morrill NE 21 F4
Morrilton AR 51 E3
Morris AL 53 D4
Morris IL 25 E4
Morris MN 14 A4
Morris OK 50 C2
Morrison IL 25 D4
Morris Plains NJ 69 E1
Morristown NJ 69 E1
Morristown TN 53 F1
Morrisville NC 55 D2
Morrisville VT 29 E1
Morro Bay CA 44 A2
Morton IL 39 F1
Morton MS 64 A1
Morton TX 48 C4
Morton Grove IL 98 D3
Moscow ID 9 E3
Moscow Mills MO 39 E3
Moses Lake WA 9 D2
Mosheim TN 54 A1
Mosinee WI 25 D1
Moss Bluff LA 63 D3
Moss Pt. MS 64 B3
Mott ND 12 C3
Moulton AL 52 C3
Moultrie GA 65 F3
Mound MN 24 B1
Mound Bayou MS 52 A4
Moundridge KS 37 F3
Mounds OK 50 B2
Mounds View MN 110 C1
Moundsville WV 41 F2
Moundville AL 64 C1
Mountainair NM 47 F3
Mountainaire AZ 46 B3
Mtn. Brook AL 96 B2
Mtn. City TN 54 B1
Mtn. Grove MO 51 E1
Mtn. Home AR 51 E1
Mtn. Home ID 19 D3
Mtn. Home NC 54 A2
Mtn. Iron MN 14 C2
Mtn. Lake MN 24 A2
Mtn. View AR 51 E2
Mtn. View CA 122 D5
Mtn. View HI 73 F3
Mtn. View MO 51 F1
Mtn. View WY 20 B4
Mtn. Vil. AK 74 B2
Mt. Airy MD 68 A3
Mt. Airy NC 54 C1
Mt. Angel OR 17 E1
Mt. Ayr IA 38 B1
Mt. Carmel IL 40 A3
Mt. Carmel PA 68 B1
Mt. Clemens MI 26 C3
Mt. Dora FL 67 E1
Mt. Gay WV 41 E4
Mt. Gilead OH 41 E1
Mt. Holly NJ 69 D2
Mt. Holly NC 54 B2
Mt. Hope WV 41 F4
Mt. Horeb WI 25 D3
Mt. Jackson VA 42 B3
Mt. Joy PA 68 B2
Mt. Juliet TN 53 D2
Mt. Kisco NY 70 B4
Mountlake Terrace WA 123 B2
Mt. Lebanon PA 117 F3
Mt. Morris IL 25 E4
Mt. Morris MI 26 C3
Mt. Morris NY 28 A3
Mt. Olive NC 55 D3
Mt. Pleasant IA 39 D1
Mt. Pleasant MI 26 B2
Mt. Pleasant PA 42 A1
Mt. Pleasant SC 66 C1
Mt. Pleasant TN 52 C2
Mt. Pleasant TX 62 C1
Mt. Pleasant UT 34 B2
Mt. Pocono PA 43 D1
Mt. Prospect IL 98 C3
Mt. Shasta CA 17 E4
Mt. Sterling KY 41 D4
Mt. Union PA 42 B1
Mt. Vernon GA 66 A2
Mt. Vernon IL 39 F3
Mt. Vernon IN 40 A4
Mt. Vernon IA 24 C4
Mt. Vernon KY 41 D4
Mt. Vernon MO 51 D1
Mt. Vernon NY 27 F3
Mt. Vernon NY 69 E1
Mt. Vernon OH 41 E1
Mt. Vernon TX 62 C1
Mt. Vernon WA 8 B1
Mt. Washington KY 40 C4
Moville IA 23 F3
Moyie Sprs. ID 9 F1
Muenster TX 50 A4
Mukilteo WA 8 B2
Mulberry AR 51 D2
Mulberry FL 67 E2
Mulberry NC 54 B1
Muldrow OK 51 D2
Muleshoe TX 48 C3
Mullan ID 9 F2
Mullens WV 41 F4
Mullins SC 55 D4
Mulvane KS 37 F4
Muncie IN 40 C1
Muncy PA 28 B4
Munday TX 49 E4
Mundelein IL 98 B1
Munds Park AZ 46 B3
Munford AL 53 D4
Munford TN 52 A2
Munfordville KY 40 B4
Munhall PA 117 G3
Munising MI 16 A3
Munster IN 98 E6
Murdo SD 22 C2
Murfreesboro AR 51 D4
Murfreesboro NC 55 E1
Murfreesboro TN 53 D2
Murphy MO 120 A3
Murphysboro IL 39 F4
Murray KY 52 C1
Murray UT 34 A1
Murrells Inlet SC 55 D4
Murrieta CA 45 D4
Muscatine IA 24 C4
Muscle Shoals AL 52 C3
Muskego WI 25 E3
Muskegon MI 26 A2
Muskegon Hts. MI 26 A2
Muskogee OK 50 C2
Mustang OK 50 A2
Myerstown PA 68 B2
Myrtle Beach SC 55 D4
Myrtle Creek OR 17 E3
Myrtle Pt. OR 17 D3
Mystic CT 71 D3
Mystic Island NJ 69 E3

N

Nacogdoches TX 62 C2
Nags Head NC 55 F2
Naknek AK 74 B3
Nampa ID 19 D2
Nanakuli HI 72 A3
Nanticoke PA 28 B4
Nantucket MA 71 F4
Nanty Glo PA 42 B1
Napa CA 31 E3
Naperville IL 25 E4
Naples FL 67 E3
Naples TX 50 C4
Naples UT 34 C1
Naples Manor FL 67 E4
Naples Park FL 67 E3
Napoleon ND 13 E3
Napoleon OH 26 B4
Nappanee IN 26 A4
Naranja FL 67 F4
Narragansett Pier RI 71 D3
Narrows VA 41 F4
Nashua IA 24 B3
Nashua NH 71 D1
Nashville AR 51 D4
Nashville GA 65 F2
Nashville IL 39 F3
Nashville NC 55 D2
Nashville TN 53 D2
Natalbany LA 63 F3
Natalia TX 59 F4
Natchez MS 63 F2
Natchitoches LA 63 D2
Natick MA 71 D2
National City CA 121 E2
Naugatuck CT 70 B3
Navajo NM 47 D2
Navasota TX 62 B3
Nazareth PA 69 D1
Nebraska City NE 38 A1
Neche ND 13 F1
Nederland CO 35 F2
Nederland TX 63 D4
Needham MA 71 E2
Needles CA 45 F3
Needville TX 61 E1
Neenah WI 25 E1
Negaunee MI 16 A3
Neillsville WI 25 D1
Nekoosa WI 25 D1
Neligh NE 23 E3
Nelsonville OH 41 E2
Neodesha KS 38 A4
Neosho MO 51 D1
Nephi UT 34 A2
Neptune Beach FL 66 B3
Neptune City NJ 69 E2
Nesquehoning PA 68 C1
Ness City KS 37 D3
Nettleton MS 52 B4
Nevada IA 24 B4
Nevada MO 38 B4
Nevada City CA 31 F2
New Albany IN 40 B3
New Albany MS 52 B3
Newark AR 51 F2
Newark CA 122 D5
Newark DE 68 C3
Newark NJ 69 E1
Newark NY 28 B3
Newark OH 41 E2
New Baden IL 39 F3
New Baltimore MI 26 C3
New Bedford MA 71 E3
Newberg OR 8 A4
New Berlin WI 25 E3
New Bern NC 55 E3
Newbern TN 52 B2
Newberry FL 66 A4
Newberry MI 16 B3
Newberry SC 54 B3
New Boston TX 51 D4
New Braunfels TX 59 F4
New Bremen OH 41 D1
New Brighton MN 110 C1
New Britain CT 70 C3
New Brunswick NJ 69 E1
New Buffalo MI 25 F4
Newburgh IN 40 A4
Newburgh NY 70 A3
Newburyport MA 71 E1
New Canaan CT 70 B4
New Carlisle OH 41 D2
New Carrollton MD 124 E1
New Castle CO 35 D2
New Castle DE 68 C3
New Castle IN 40 C2
Newcastle OK 50 A2
New Castle PA 41 F1
Newcastle WY 21 F2
New City NY 70 A4
Newcomerstown OH 41 E1
New Concord OH 41 E2
Newell SD 22 A1
New Ellenton SC 54 B4
New England ND 12 C3
New Fairfield CT 70 B3
Newfane NY 27 F2
New Freedom PA 68 B3
New Glarus WI 25 D3
New Hampton IA 24 C3
New Haven CT 70 C4
New Haven IN 40 C1
New Haven MI 26 C3
New Haven MO 39 D3
New Haven WV 41 E3
New Holland PA 68 C2
New Holstein WI 25 E2
New Hope AL 53 D3
New Hope MN 110 B2
New Hope MS 52 C4
New Hyde Park NY 113 G3
New Iberia LA 63 E4
Newington CT 70 C3
New Johnsonville TN 52 C2
New Kensington PA 42 A1
Newkirk OK 50 B1
New Lebanon OH 41 D2
New Lenox IL 98 B6
New Lexington OH 41 E2
New Llano LA 63 D3
New London CT 71 D3
New London IA 39 D1
New London OH 41 E1
New London WI 25 E1
New Madrid MO 52 B1
Newman CA 31 F4
New Market AL 53 D3
Newmarket NH 29 F3
New Market VA 42 B3
New Martinsville WV 41 F2
New Meadows ID 19 D1
New Milford CT 70 B3
New Milford NJ 112 C1
Newnan GA 53 E4
New Orleans LA 64 A4
New Paltz NY 70 A3
New Philadelphia OH 41 F1
New Plymouth ID 19 D2
Newport AR 51 F2
Newport KY 99 B3
Newport ME 30 B3
Newport NH 29 E2
Newport NC 55 E3
Newport OR 17 D1
Newport RI 71 E3
Newport TN 53 F2
Newport VT 29 E1
Newport WA 9 E1
Newport Beach CA 44 C4
Newport News VA 55 F1
New Port Richey FL 67 D1
New Prague MN 24 B1
New Richmond WI 14 C4
New River AZ 46 B4
New Roads LA 63 F3
New Rochelle NY 69 E1
New Rockford ND 13 E2
New Salem ND 13 D3
New Smyrna Beach FL 67 E1
New Tazewell TN 53 F1
Newton AL 65 D2
Newton IL 40 A3
Newton IA 24 B4
Newton KS 37 F4
Newton MA 71 E2
Newton MS 64 B1
Newton NJ 43 E1
Newton NC 54 B2
Newton TX 63 D3
New Town ND 12 C2
New Ulm MN 24 A1
New Underwood SD 22 A2
New Whiteland IN 40 B2
New Windsor NY 70 A3
New York NY 69 E1
Nezperce ID 9 F3
Niagara WI 15 F4
Niagara Falls NY 27 F3
Niantic CT 70 C3
Niceville FL 65 D3
Nicholasville KY 40 C4
Nicholson MS 64 A3
Nickerson KS 37 E3
Nikiski AK 74 C3
Niles IL 98 D3
Niles MI 26 A4
Niles OH 27 E4
Ninety Six SC 54 A4
Nipomo CA 44 A3
Nisswa MN 14 B3
Nitro WV 41 E3
Niwot CO 35 F2
Nixa MO 51 E1
Nixon NV 32 B2
Nixon TX 59 F4
Noble OK 50 A3
Noblesville IN 40 B2
Nocona TX 50 A4
Noel MO 51 D1
Nogales AZ 56 B2
Nolanville TX 62 A3
Nolensville TN 53 D2
Nome AK 74 B2
Noorvik AK 74 B2
Nora Sprs. IA 24 B2
Norco CA 45 D4
Norfolk NE 23 E4
Norfolk VA 55 F1
Normal IL 39 F1
Norman OK 50 A2
Norridgewock ME 29 F1
Norristown PA 68 C2
N. Adams MA 70 B1
N. Amherst MA 70 C2
Northampton MA 70 C2
Northampton PA 68 C1
N. Arlington NJ 112 B3
N. Atlanta GA 94 C1
N. Attleboro MA 71 E2
N. Augusta SC 54 A4
N. Baltimore OH 26 C4
N. Bend NE 23 E4
N. Bend OR 17 D2
N. Bend WA 8 B2
N. Bennington VT 70 B1
N. Bergen NJ 112 C2
N. Berwick ME 29 F2
Northborough MA 71 D2
N. Branch MN 14 C4
Northbridge MA 71 D2
Northbrook IL 98 C2
Northbrook OH 99 A1
N. Brunswick NJ 69 E1
N. Canton OH 41 F1
N. Cape May NJ 69 D4
N. Charleston SC 66 C1
N. Chicago IL 25 F3
N. College Hill OH 99 A2
N. Conway NH 29 F2
N. Crossett AR 51 F4
Northdale FL 123 E1
N. Decatur GA 94 D2
N. Druid Hills GA 94 C2
North East MD 68 C3
N. East PA 27 E3
N. Fair Oaks CA 122 C5
Northfield MN 24 B1
Northfield NJ 69 D3
Northfield VT 29 E1
N. Fond du Lac WI 25 E2
N. Ft. Myers FL 67 E3
Northglenn CO 101 B1
N. Haven CT 70 C3
N. Highlands CA 31 F3
N. Kingsville OH 27 E4
N. Las Vegas NV 45 E1
N. Lauderdale FL 109 A2
N. Liberty IA 24 C4
N. Little Rock AR 51 E3
N. Manchester IN 40 C1
N. Mankato MN 24 A1
N. Miami FL 67 F4
N. Miami Beach FL 67 F4
N. Muskegon MI 26 A2
N. Myrtle Beach SC 55 D4
N. Naples FL 67 E3
N. Ogden UT 20 A4
N. Olmsted OH 27 D4
N. Palm Beach FL 67 F3
N. Plainfield NJ 69 E1
N. Platte NE 22 C4
N. Pole AK 74 C2

Northport AL 52 C4	N. Salt Lake UT 34 A1	N. Vernon IN 40 C3	Norton Shores MI 26 A2	Norwich NY 28 C3	Nowata OK 50 C1	Oceano CA 44 A3	Olivehurst CA 31 F2
N. Port FL 67 D3	N. Sioux City SD 23 F3	N. Wildwood NJ 69 D4	Norwalk CA 106 D3	Norwood MA 71 E2	Nowata OK 50 C1	Ocean Shores WA 8 A3	Oliver Sprs. TN 53 F2
Northport NY 69 F1	N. Springfield VA 124 A3	N. Wilkesboro NC 54 B2	Norwalk CT 70 B4	Norwood NC 54 C2	Nutley NJ 112 B2	Oceanside CA 56 A3	Olivia MN 24 A1
N. Providence RI 71 D3	N. Syracuse NY 28 B2	N. Windham ME 29 F2	Norwalk IA 24 B4	Norwood OH 99 B2	Nyack NY 70 A4	Oceanside NY 69 F1	Olney IL 40 A3
N. Richland Hills TX 100 D2	N. Terre Haute IN 40 A2	Northwood IA 24 B2	Norwalk OH 26 C4	Norwood-Young America	Nyssa OR 18 C2	Ocean Sprs. MS 64 B3	Olney MD 68 A4
Northridge OH 41 D2	N. Tonawanda NY 27 F3	Northwood ND 13 F2	Norway ME 29 F1	MN 24 B1		Ocilla GA 65 F2	Olney TX 49 F4
N. Royalton OH 99 E3	Northumberland PA 68 B1	Norton KS 37 D2	Norway MI 15 F4	Novato CA 31 E3	**O**	Ocoee FL 67 E1	Olton TX 48 C4
N. St. Paul MN 110 E2	N. Valley Stream NY 113 G4	Norton VA 54 A1	Norwich CT 71 D3	Novi MI 101 E1	Oak Creek WI 25 F3	Oconomowoc WI 25 E3	Olympia WA 8 B3
					Oakdale CA 31 F4		Omaha NE 23 F4

Oakdale LA 63 E3	Oconto WI 25 F1	Osage IA 24 B2
Oakdale MN 110 E2	Oconto Falls WI 25 E1	Osage Beach MO 38 C4
Oakes ND 13 F3	Odem TX 61 D2	Osage City KS 38 A3
Oak Forest IL 98 D6	Odenton MD 68 B4	Osakis MN 14 A4
Oak Grove KY 52 C4	Odessa FL 67 D1	Osawatomie KS 38 B3
Oak Grove LA 63 F1	Odessa MO 38 C3	Osborne KS 37 F2
Oak Grove MN 14 C4	Odessa TX 58 C2	Osburn ID 9 F2
Oak Grove MS 64 A2	Oelwein IA 24 C3	Osceola AR 52 A2
Oak Grove MO 38 B3	O'Fallon IL 39 E3	Osceola IA 38 C4
Oak Grove OR 118 B3	O'Fallon MO 39 E3	Osceola IN 23 E4
Oak Harbor OH 26 C4	Ogallala NE 22 B4	Osceola WI 14 C4
Oak Harbor WA 8 B1	Ogden IA 23 A4	Oshkosh NE 22 B4
Oak Hill WV 41 F4	Ogden KS 37 F2	Oshkosh WI 25 E2
Oakhurst CA 31 F4	Ogden NC 55 E3	Oskaloosa IA 24 B4
Oak Island NC 55 D4	Ogden UT 20 A4	Oskaloosa KS 38 B2
Oakland CA 31 E4	Ogdensburg NY 28 C1	Osprey FL 67 D2
Oakland IA 23 F4	Oglala SD 22 A4	Osseo MN 14 C4
Oakland ME 29 F1	Oglesby IL 25 E4	Osseo WI 24 C1
Oakland NE 23 F4	Oil City PA 27 E4	Ossian IN 40 C1
Oakland City IN 40 A3	Oildale CA 44 B2	Ossining NY 70 A4
Oakland Park FL 109 B2	Ojai CA 44 B3	Oswego IL 25 E4
Oak Lawn IL 25 F4	Okanogan WA 9 D1	Oswego KS 38 B4
Oakley ID 19 F3	Okarche OK 50 A2	Oswego NY 28 B2
Oakley KS 36 C3	Okeechobee FL 67 F2	Othello WA 9 D3
Oak Park IL 25 F4	Okeene OK 49 F2	Otsego MI 26 A3
Oak Park MI 101 F1	Okemah OK 50 B2	Ottawa IL 25 E4
Oakridge OR 17 E2	Oklahoma City OK 50 A2	Ottawa KS 38 B3
Oak Ridge TN 53 F2	Okmulgee OK 50 C2	Ottawa OH 41 D1
Oak Ridge North TX 62 B4	Okolona MS 52 B4	Ottumwa IA 39 D1
Oakton VA 68 A4	Ola AR 51 E3	Overgaard AZ 46 C3
Oak View CA 44 B3	Olathe CO 35 D3	Overland MO 120 B3
Oakville CT 70 B3	Olathe KS 38 B3	Overland Park KS 38 B3
Oakville MO 120 B3	Old Bridge NJ 69 E2	Overlea MD 95 C2
Oakwood GA 53 F3	Old Forge PA 28 B4	Overton NV 45 F1
Oberlin KS 37 D2	Old Orchard Beach	Overton TX 62 C1
Oberlin LA 63 E3	ME 29 F2	Oviedo FL 67 E1
Oberlin OH 27 D4	Oldsmar FL 67 D1	Ovilla TX 62 A1
Ocala FL 66 A4	Old Town ME 30 B3	Owasso OK 50 C1
Oceana WV 41 F4	Olean NY 27 F4	Owatonna MN 24 B2
Ocean City MD 43 D2	Olive Branch MS 52 A3	Owego NY 28 B4
Ocean City NJ 69 D4	Olive Hill KY 41 D3	Owensboro KY 40 B4
		Owensville MO 39 D3
		Owenton KY 40 C3
		Owings Mills MD 68 B3

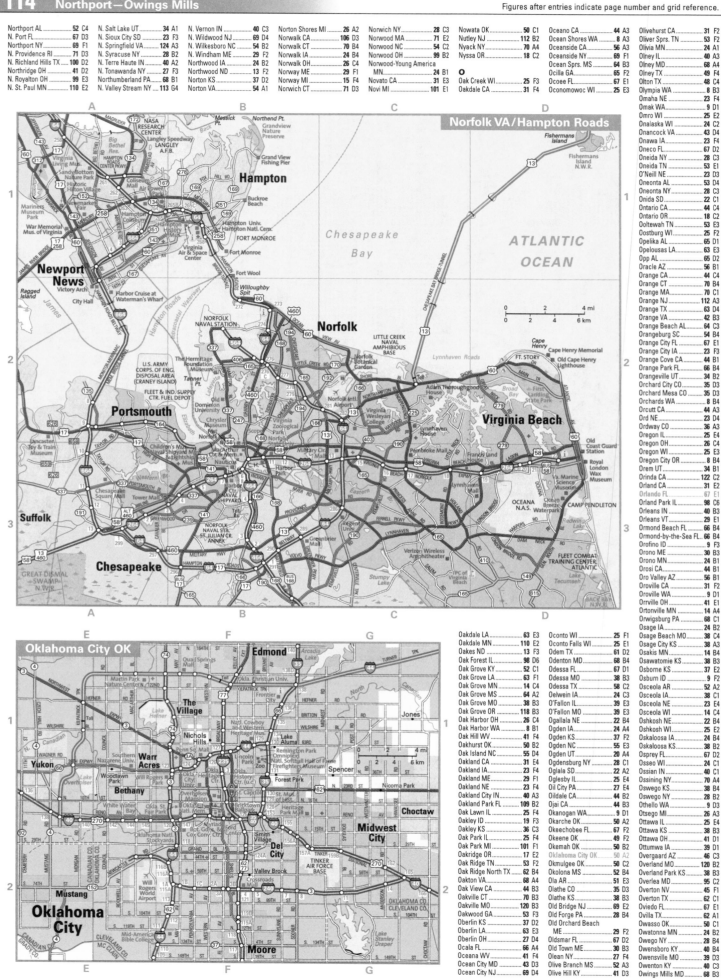

Norfolk VA/Hampton Roads

Oklahoma City OK

Entries in **bold color** indicate cities with detailed inset maps.
Owingsville—Portsmouth 115

Owingsville KY 41 D4
Owosso MI 26 B3
Owyhee NV 19 D4
Oxford AL 53 D4
Oxford KS 37 F4
Oxford ME 29 F1
Oxford MA 71 D2
Oxford MI 26 C3
Oxford MS 52 B3
Oxford NE 37 D1
Oxford NC 55 D1
Oxford PA 68 C3
Oxnard CA 44 B4
Oxon Hill MD 68 A4
Oyster Bay NY 69 F1
Ozark AL 65 D2
Ozark AR 51 D2
Ozark MO 51 E1
Ozona TX 59 D3

P
Pablo MT 10 A2
Pace FL 64 C3
Pacific MO 39 E3
Pacifica CA 31 E4
Pacific Grove CA 31 D4
Pacolet SC 54 B3
Paddock Lake WI 25 E3
Paden City WV 41 F2
Paducah KY 52 B1
Paducah TX 49 E4
Page AZ 46 B1
Pageland SC 54 C3
Pagosa Sprs. CO 35 E4
Pahokee FL 67 F3
Pahrump NV 45 E2
Paia HI 73 D1
Painesville OH 27 D4
Paintsville KY 41 E4
Palacios TX 61 E2
Palatine IL 25 F3
Palatka FL 66 B4
Palermo CA 31 F2
Palestine TX 62 B2
Palisade CO 35 D3
Palisades Park NJ 112 D2
Palm Bay FL 67 F2
Palm Beach FL 67 F3
Palm Beach Gardens FL 67 F3
Palm Coast FL 66 B4
Palmdale CA 44 C3
Palm Desert CA 45 D4
Palmer AK 74 C3
Palmer MA 70 C2
Palmer TX 62 A1
Palmer Lake CO 35 F2
Palmerton PA 68 C1
Palmetto FL 67 D2
Palmetto GA 53 D4
Palmetto Bay FL 109 A5
Palm Harbor FL 67 D2
Palm River FL 123 E2
Palm Sprs. CA 45 D4
Palm Valley FL 66 B3
Palmyra MO 39 D2
Palmyra NJ 69 D2
Palmyra NY 28 A3
Palmyra PA 68 B2
Palmyra WI 25 E3
Palo Alto CA 31 F4
Palos Hills IL 98 C5
Pampa TX 49 D2
Pana IL 39 F2
Panaca NV 33 E4
Panama OK 50 C3
Panama City FL 65 D4
Panama City Beach FL 65 D4
Panguitch UT 34 A4
Panhandle TX 49 D2
Panthersville GA 94 D3
Paola KS 38 B3
Paoli IN 40 B3
Paoli PA 68 C2
Paonia CO 35 D3
Papillion NE 23 F4
Parachute CO 35 D2
Paradise CA 31 F2
Paradise NV 105 B2
Paradise Valley AZ 46 B4
Paragould AR 52 A2
Paramount CA 106 D3
Paramus NJ 112 C1
Pardeeville WI 25 E2
Paris AR 51 D2
Paris ID 20 A3
Paris IL 40 A2
Paris KY 41 D3
Paris MO 39 D2
Paris TN 52 C1
Paris TX 50 C4
Park City KS 37 F4
Park City MT 11 E4
Park City UT 34 B1
Parker AZ 45 F4
Parker CO 35 F2
Parker FL 65 D4
Parker SD 23 E2
Parkersburg IA 24 B3
Parkersburg WV 41 F2
Parkesburg PA 68 C2
Park Falls WI 15 E4
Park Hill OK 50 C2
Park Hills MO 39 E4
Parkin AR 52 A2
Parkland FL 109 A1
Parkland WA 8 B2
Park Rapids MN 14 B3
Park Ridge IL 98 C3
Park River ND 13 F1
Parks AZ 46 B2
Parkston SD 23 E2
Parkville MD 95 C1
Parlier CA 44 B1
Parma ID 18 C2
Parma OH 27 D4
Parma Hts. OH 99 E3
Parmelee SD 22 C2
Parowan UT 34 A4
Parshall ND 12 C2
Parsons KS 38 B4
Parsons TN 52 C2
Parsons WV 42 A3
Pasadena CA 44 C4
Pasadena TX 61 F1
Pascagoula MS 64 B3
Pasco WA 9 D3
Pascoag RI 71 D2
Paso Robles CA 44 A2
Passaic NJ 69 E1
Pass Christian MS 64 A3
Patchogue NY 69 F1
Paterson NJ 69 E1
Patrick Sprs. VA 54 C1
Patterson CA 31 F4
Patterson LA 63 F4
Paul ID 19 F3
Paulden AZ 46 B3
Paulding OH 26 B4
Pauls Valley OK 50 B3
Pauwela HI 73 D1
Pawcatuck CT 71 D3
Pawhuska OK 50 B1
Pawnee IL 39 F2
Pawnee OK 50 B1
Pawnee City NE 38 A2
Paw Paw MI 26 A3
Pawtucket RI 71 D2
Paxton IL 40 A1
Payette ID 18 C2
Paynesville MN 14 B4
Payson AZ 46 B3
Payson UT 34 B2
Peabody KS 37 F3
Peabody MA 71 E1
Peachtree City GA 53 E4
Pea Ridge AR 51 D1
Pearisburg VA 41 F4
Pearl MS 64 A1
Pearland TX 61 F1
Pearl City HI 72 A3
Pearlington MS 64 A3
Pearl River LA 64 A3
Pearsall TX 60 C1
Pecos NM 48 A2
Pecos TX 58 B2
Peculiar MO 38 B3
Peekskill NY 70 A4
Pegram TN 53 D2
Pekin IL 39 F1
Pelahatchie MS 64 A1
Pelham AL 53 D4
Pelham GA 65 E3
Pelican Rapids MN 14 A3
Pella IA 24 B4
Pell City AL 53 D4
Pembina ND 13 F1
Pembroke GA 66 B2
Pembroke NC 55 D3
Pembroke Pines FL 109 A3
Pena Blanca NM 47 F2
Pen Argyl PA 69 D1
Penasco NM 48 A1
Pender NE 23 E3
Pendleton IN 40 C2
Pendleton OR 9 D4
Pendleton SC 54 A3
Penn Hills PA 117 H1
Pennington Gap VA 53 F1
Pennsauken NJ 116 D3
Pennsboro WV 41 F2
Pennsburg PA 68 C2
Penns Grove NJ 43 D2
Pennsville NJ 68 C3
Penn Yan NY 28 B3
Penrose CO 35 F3
Pensacola FL 64 C3
Peoria AZ 46 B4
Peoria IL 39 F1
Peotone IL 25 F4
Pepeekeo HI 73 F3
Peralta NM 47 F3
Perham MN 14 B3
Peridot AZ 46 C4
Perkasie PA 69 D2
Perkins OK 50 B2
Perrine FL 67 F4
Perris CA 45 D4
Perry FL 65 F4
Perry GA 65 F1
Perry IA 24 A4
Perry MI 26 B3
Perry NY 27 F3
Perry OK 50 B1
Perry UT 20 A4
Perry Hall MD 68 B3
Perrysburg OH 26 C4
Perryton TX 49 D1
Perryville AR 51 E3
Perryville MD 68 C3
Perryville MO 39 E4
Perth Amboy NJ 69 E1
Peru IL 25 E4
Peru IN 40 B1
Peshtigo WI 16 A4
Petal MS 64 A2
Petaluma CA 31 E3
Peterborough NH 71 D1
Petersburg AK 75 E4
Petersburg IN 40 A3
Petersburg VA 42 C4
Petersburg WV 42 A3
Petersville AL 52 C3
Petoskey MI 16 B4
Pevely MO 39 E3
Pflugerville TX 62 A3
Pharr TX 60 C4
Phenix City AL 65 E1
Philadelphia MS 64 B1
Philadelphia PA 69 D2
Philip SD 22 B2
Philippi WV 42 A2
Philipsburg MT 10 B3
Philipsburg PA 42 B1
Phillips WI 15 E4
Phillipsburg KS 37 D2
Phillipsburg NJ 69 D1
Philomath OR 17 E1
Phoenix AZ 46 B4
Phoenix OR 17 E3
Phoenixville PA 68 C2
Picayune MS 64 A3
Picher OK 50 C1
Pickens MS 64 A1
Pickens SC 54 A3
Pickerington OH 41 E2
Pico Rivera CA 106 D3
Piedmont AL 53 E4
Piedmont MO 51 F1
Piedmont OK 50 A2
Piedmont SC 54 A3
Piedmont SD 22 A1
Pierce ID 9 F3
Pierce NE 23 E3
Pierre SD 22 C1
Pierre Part LA 63 F4
Pigeon Forge TN 53 F2
Piggott AR 52 A1
Pikesville MD 95 A1
Pikeville KY 41 E4
Pikeville TN 53 E2
Pilot Pt. TX 50 B4
Pilot Rock OR 9 D4
Pilot Sta. AK 74 B3
Pima AZ 56 C1
Pinckney MI 26 B3
Pinckneyville IL 39 F4
Pine AZ 46 B3
Pine Bluff AR 51 F3
Pine Bluffs WY 21 F4
Pine City MN 14 C4
Pinecrest FL 109 A5
Pinedale WY 20 B3
Pine Hills FL 115 B2
Pinehurst ID 9 F2
Pinehurst NC 54 C3
Pinehurst TX 62 B4
Pine Island MN 24 B1
Pine Knot KY 53 E1
Pinellas Park FL 67 D2
Pine Ridge SD 22 B3
Pinesdale MT 10 A3
Pinetop-Lakeside AZ 46 C4
Pineville KY 53 F1
Pineville LA 63 E2
Piney Green NC 55 E3
Pink OK 50 B2
Pinole CA 122 C1
Pinon AZ 46 C2
Pinson AL 53 D4
Pioche NV 33 E4
Pipestone MN 23 F2
Piqua OH 41 D2
Pirtleville AZ 56 C2
Pismo Beach CA 44 A3
Pitman NJ 69 D3
Pittsboro NC 55 D2
Pittsburg CA 31 F4
Pittsburg KS 38 B4
Pittsburg TX 62 C1
Pittsburgh PA 42 A1
Pittsfield IL 39 E2
Pittsfield ME 30 B3
Pittsfield MA 70 B1
Pittsfield NH 29 E2
Pittston PA 28 B4
Placentia CA 107 F4
Placerville CA 31 F3
Placitas NM 47 F2
Plain City OH 41 D2
Plain City UT 20 A4
Plainfield CT 71 D3
Plainfield IL 25 E4
Plainfield IN 40 B2
Plainfield NJ 69 E1
Plainfield VT 29 E1
Plains KS 37 D4
Plains MT 10 A2
Plains TX 58 B1
Plainview MN 24 C1
Plainview NE 23 E3
Plainview TX 49 D4
Plainville KS 37 E2
Plainville MA 71 E2
Plainwell MI 26 A3
Plaistow NH 71 E1
Planada CA 32 A4
Plankinton SD 23 D2
Plano IL 25 E4
Plano TX 62 A1
Plantation FL 67 F3
Plant City FL 67 D2
Plantersville MS 52 B4
Plaquemine LA 63 F3
Platte SD 23 D2
Platte City MO 38 B2
Platteville CO 35 F1
Platteville WI 25 D3
Plattsburg MO 38 B2
Plattsburgh NY 29 D1
Plattsmouth NE 38 A1
Pleasant Garden NC 54 C2
Pleasant Grove AL 96 A1
Pleasant Grove UT 34 B1
Pleasant Hill CA 122 D2
Pleasant Hill MO 38 B3
Pleasant Hills MD 68 B3
Pleasanton CA 31 F4
Pleasanton KS 38 B3
Pleasanton TX 60 C1
Pleasant Prairie WI 25 F3
Pleasant View TN 53 D1
Pleasant View UT 20 A4
Pleasantville IA 24 B4
Pleasantville NJ 69 E3
Pleasantville NY 70 A4
Plentywood MT 12 B1
Plover WI 25 D1
Plummer ID 9 E2
Plymouth IN 26 A4
Plymouth MA 71 E2
Plymouth MI 101 E2
Plymouth MN 110 A2
Plymouth NH 29 E2
Plymouth NC 55 E2
Plymouth WI 25 F2
Pocahontas AR 51 F1
Pocahontas IA 24 A3
Pocatello ID 20 A3
Pocola OK 51 D2
Pocomoke City MD 43 D4
Pt. Clear AL 64 B3
Pt. Hope AK 74 B1
Pt. Pleasant NJ 69 E2
Pt. Pleasant WV 41 E3
Pt. Pleasant Beach NJ 69 E2
Pojoaque NM 47 F2
Polacca AZ 46 C2
Polk City IA 24 B4
Pollock Pines CA 32 A3
Polson MT 10 A2
Pomeroy WA 9 E3
Pomona CA 44 C4
Pomona NJ 69 D3
Pompano Beach FL 67 F3
Pompton Lakes NJ 70 A4
Ponca NE 23 E3
Ponca City OK 50 B1
Ponchatoula LA 63 F3
Pontiac IL 39 F1
Pontiac MI 26 C3
Pontotoc MS 52 B3
Pooler GA 66 B2
Poolesville MD 68 A4
Poplar MT 12 A1
Poplar Bluff MO 52 A1
Poplarville MS 64 A3
Poquoson VA 55 F1
Portage IN 25 F4
Portage MI 26 A3
Portage PA 42 B1
Portage WI 25 E2
Portageville MO 52 B1
Portales NM 48 C4
Port Allen LA 63 F3
Port Angeles WA 8 A2
Port Aransas TX 61 D3
Port Arthur TX 63 D4
Port Barre LA 63 E3
Port Charlotte FL 67 D3
Port Chester NY 70 B4
Port Clinton OH 26 C4
Port Edwards WI 25 D1
Porterville CA 44 B2
Port Ewen NY 70 A2
Port Gibson MS 63 F2
Port Hadlock WA 8 B2
Port Hueneme CA 44 B4
Port Huron MI 26 C3
Port Isabel TX 61 D4
Port Jefferson NY 69 F1
Port Jervis NY 28 C4
Portland CT 70 C3
Portland IN 40 C1
Portland ME 29 F2
Portland MI 26 B3
Portland ND 13 F2
Portland OR 8 B4
Portland TN 53 D1
Portland TX 61 D2
Port Lavaca TX 61 E2
Port Ludlow WA 8 B2
Port Monmouth NJ 69 E1
Port Neches TX 63 D4
Port Orange FL 67 E1
Port Orchard WA 8 B2
Port Richey FL 67 D1
Port Royal SC 66 B1
Port St. Joe FL 65 E4
Port St. John FL 67 F1
Port St. Lucie FL 67 F2
Port Salerno FL 67 F2
Portsmouth NH 29 F3
Portsmouth OH 41 E3
Portsmouth RI 71 E3
Portsmouth VA 55 F1

Port Sulphur LA 64 A4
Port Townsend WA 8 B2
Port Washington NY ... 113 G2
Port Washington WI 25 F2
Port Wentworth GA 66 B2
Post TX 59 D1
Post Falls ID 9 E2
Postville IA 24 C3
Poteau OK 51 D3
Poteet TX 60 C1
Poth TX 60 C1
Potlatch ID 9 E3
Potomac MD 68 A4
Potosi MO 39 E4
Potosi TX 59 E1
Potsdam NY 28 C1
Potterville MI 26 B3
Pottsboro TX 50 B4
Pottstown PA 68 C2
Pottsville AR 51 E2
Pottsville PA 68 B1
Poughkeepsie NY 70 A3
Poulsbo WA 8 B2
Poultney VT 29 E1
Poway CA 56 B3
Powder Spgs. GA 53 E4
Powell TN 53 F2
Powell WY 20 C1
Poynette WI 25 E2
Prague OK 50 B2
Prairie du Chien WI ... 24 C3
Prairie du Sac WI 25 D2
Prairie Grove AR 51 D2
Prairie View TX 62 B4
Prairie Vil. KS 104 B4
Pratt KS 37 E4
Prattville AL 65 D1
Premont TX 60 C3
Prentiss MS 64 A2
Prescott AZ 46 B3
Prescott AR 51 D4
Prescott WI 24 B1
Prescott Valley AZ 46 B3
Presho SD 22 C2
Presidio TX 58 A4
Presque Isle ME 30 C1
Preston ID 20 A3
Preston MN 24 C2

Prestonsburg KY 41 E4
Price UT 34 B2
Priceville AL 53 D3
Prichard AL 64 B3
Priest River ID 9 E1
Princeton IL 25 E4
Princeton IN 40 A3
Princeton KY 52 C1
Princeton MN 14 C4
Princeton NJ 69 D2
Princeton TX 62 B1
Princeton WV 41 F4
Princeton WI 25 E2
Princeville HI 72 B1
Princeville OR 17 F1
Prineville OR 17 F1
Prior Lake MN 24 B1
Proctor MN 15 D3
Proctor VT 29 D2
Progreso TX 60 C4
Prospect Hts. IL 98 C2
Prosper TX 50 B4
Prosser WA 9 D3
Providence KY 40 A4
Providence RI 71 D3
Providence UT 20 A4
Provincetown MA 71 F2
Provo UT 34 B1
Prunedale CA 44 A1
Pryor MT 11 E4
Pryor OK 50 C1
Pueblo CO 36 A3
Pueblo West CO 36 A3
Pukalani HI 72 B2
Pulaski TN 53 D3
Pulaski VA 54 B1
Pullman WA 9 E3
Punta Gorda FL 67 D2
Punxsutawney PA 42 B1
Pupukea HI 72 A2
Purcell OK 50 A3
Purcellville VA 42 B3
Purvis MS 64 A3
Putnam CT 71 D2
Putnam Lake NY 70 B3
Putney GA 65 F2
Putney VT 70 C1
Puyallup WA 8 B2

Q
Quakertown PA 68 C1
Quanah TX 49 E3
Quartz Hill CA 44 C3
Quartzsite AZ 45 F4
Queen City TX 62 C1
Queen Creek AZ 46 B4
Questa NM 48 A1
Quince Orchard MD 68 A4
Quincy FL 65 E3
Quincy IL 39 D2
Quincy MA 71 E1
Quincy WA 9 D2
Quinhagak AK 74 B3
Quitman GA 65 F3
Quitman MS 64 B2
Quitman TX 62 B1

R
Raceland KY 41 E3
Raceland LA 63 F4
Racine WI 25 F3
Radcliff KY 40 B4
Radford VA 41 F4
Radium Sprs. NM 57 D3
Radnor PA 116 A2
Raeford NC 55 D3
Ragland AL 53 D4
Rahway NJ 69 E1
Rainbow City AL 53 D4
Rainelle WV 41 F4
Rainier OR 8 B4
Rainsville AL 53 D3
Raleigh MS 64 A2
Raleigh NC 55 D2
Ralls TX 49 D4
Ramona CA 56 B3
Ramsey MN 14 C4
Ramsey NJ 70 A4
Ranchester WY 21 D1
Rancho Cordova CA 31 F3
Rancho Cucamonga
  CA 107 H2
Rancho Mirage CA 45 D4
Rancho Palos Verdes
  CA 106 C4
Rancho Santa Margarita
  CA 107 H5

Ranchos de Taos NM 48 A1
Randallstown MD 68 B3
Randleman NC 54 C2
Randolph MA 71 E1
Randolph NE 23 E3
Randolph VT 29 E2
Randolph WI 25 E2
Random Lake WI 25 F2
Rangely CO 35 D2
Ranger TX 59 F1
Rantoul IL 40 A1
Rapid City SD 22 A1
Raritan NJ 69 D1
Rathdrum ID 9 E2
Ratliff FL 66 A3
Raton NM 48 B1
Raven VA 54 B1
Ravena NY 70 A1
Ravenel SC 66 C1
Ravenna NE 37 E1
Ravenna OH 27 D4
Ravenswood WV 41 E3
Rawlins WY 21 D4
Ray ND 12 C1
Raymond MS 63 F1
Raymond NH 71 F3
Raymond WA 8 A3
Raymondville TX 60 C4
Raymore MO 38 B3
Rayne LA 63 E3
Raytown MO 104 C4
Rayville LA 63 E1
Reading MA 71 E1
Reading OH 99 B1
Reading PA 68 C2
Rector AR 52 A1
Red Bank NJ 69 E2
Red Bank SC 54 B4
Red Bank TN 53 E3
Red Bay AL 52 C3
Red Bluff CA 31 E1
Red Bud IL 39 E3
Red Cloud NE 37 E1
Redding CA 31 E1
Redfield SD 23 D1
Redford MI 101 F2
Richboro PA 69 D2
Red Lake MN 14 B2
Red Lake Falls MN 14 A2

Redlands CA 45 D4
Red Lion PA 68 B2
Red Lodge MT 11 E4
Redmond OR 17 F1
Redmond WA 8 B2
Red Oak IA 38 B1
Red Oak NC 55 E2
Red Oak TX 62 A1
Redondo Beach CA 106 C4
Red Sprs. NC 55 D3
Red Wing MN 24 B1
Redwood City CA 31 E4
Redwood Falls MN 23 F1
Reed City MI 26 A2
Reedley CA 44 B1
Reedsburg WI 25 D2
Reedsport OR 17 D2
Reform AL 52 C4
Refugio TX 61 D2
Rehoboth Beach DE 43 D3
Reidsville GA 66 A2
Reidsville NC 54 C1
Reinbeck IA 24 B3
Reisterstown MD 68 B3
Reliance WY 20 C4
Remsen IA 23 F3
Reno NV 32 B2
Reno TX 50 C4
Reno TX 59 F1
Rensselaer IN 40 B1
Rensselaer NY 70 B1
Renton WA 8 B2
Republic MO 51 D1
Reserve LA 63 F4
Reston VA 68 A4
Revere MA 96 F1
Rexburg ID 20 A2
Reynoldsburg OH 41 E2
Reynoldsville PA 42 B1
Rhinebeck NY 70 A2
Rhinelander WI 15 F4
Rialto CA 45 D4
Rice Lake WI 15 D4
Richardson TX 100 E1
Richardton ND 12 C3
Richboro PA 69 D2
Richfield MN 110 B3
Richfield UT 34 A3

Richford VT 29 E1
Rich Hill MO 38 B4
Richland MS 64 A1
Richland MO 39 D4
Richland WA 9 D3
Richland Ctr. WI 25 D2
Richlands VA 54 B1
Richmond CA 31 E4
Richmond IN 40 C2
Richmond KY 41 D4
Richmond ME 29 F2
Richmond MI 26 C3
Richmond MO 38 B2
Richmond TX 61 E1
Richmond UT 20 A4
Richmond VA 42 C4
Richmond Hts. MO 120 B2
Richmond Hts. OH 99 G1
Richmond Hill GA 66 B2
Richwood LA 63 E1
Richwood OH 41 D2
Richwood WV 41 F3
Ridge NY 70 C4
Ridgecrest CA 44 C2
Ridgefield CT 70 B2
Ridgefield NJ 112 D2
Ridgefield Park NJ ... 112 C1
Ridgeland MS 64 A1
Ridgeland SC 66 B1
Ridgely TN 52 B1
Ridge Manor FL 67 D1
Ridgeville SC 66 C1
Ridgewood NJ 70 A4
Ridgway PA 27 F4
Rifle CO 35 D2
Rigby ID 20 A2
Rincon GA 66 B2
Ringgold GA 53 E3
Ringgold LA 63 D1
Ringling OK 50 A3
Ringwood NJ 70 A4
Rio Bravo TX 60 B3
Rio Dell CA 31 D1
Rio Grande City TX 60 B4
Rio Hondo TX 61 D4
Rio Rancho NM 47 F2

Rio Rico AZ 56 B2
Rio Vista CA 31 F3
Ripley MS 52 B3
Ripley TN 52 B2
Ripley WV 41 E3
Ripon CA 31 F4
Ripon WI 25 E2
Ririe ID 20 A2
Rising Sun IN 40 C3
Rison AR 51 E4
Rittman OH 41 E1
Ritzville WA 9 D2
Riverbank CA 31 F4
Riverdale IL 98 E6
River Edge NJ 112 C1
River Falls WI 24 B1
Riverhead NY 70 C4
River Ridge LA 111 B2
River Rouge MI 101 G3
Riverside CA 45 D4
Riverton IL 39 F2
Riverton UT 34 A1
Riverton WY 20 C3
Riverview FL 67 D2
Riviera Beach FL 67 F3
Riviera Beach MD 68 B4
Roanoke AL 65 D1
Roanoke VA 42 A4
Roanoke Rapids NC 55 E1
Robbinsdale MN 110 B2
Roberts ID 20 A2
Robertsdale AL 64 C3
Robins IA 24 B3
Robinson IL 40 A3
Robstown TX 61 D3
Rochelle IL 25 E4
Rochester IN 40 B1
Rochester MN 24 C1
Rochester NH 29 F2
Rochester NY 28 A2
Rochester WA 8 A3
Rochester Hills MI 26 C3
Rockaway NJ 69 E1
Rockdale TX 59 F2
Rock Falls IL 25 D4

Rockford IL 25 E3
Rockford MI 26 A2
Rockford MN 14 C4
Rock Hill SC 54 B3
Rockingham NC 55 D3
Rock Island IL 24 C4
Rockland ME 30 B4
Rockland MA 71 E1
Rockledge FL 67 F1
Rocklin CA 31 F3
Rockmart GA 53 E4
Rockport IN 40 B4
Rockport MA 71 E1
Rockport TX 61 D2
Rock Rapids IA 23 F2
Rock Sprs. WY 20 C4
Rockton IL 25 E3
Rock Valley IA 23 F2
Rockville IN 40 A2
Rockville MD 68 A4
Rockwall TX 62 B1
Rockwell City IA 24 A3
Rockwood MI 26 C4
Rockwood TN 53 E2
Rocky Ford CO 36 A4
Rocky Mount NC 55 E2
Rocky Mount VA 54 C1
Rocky Pt. NY 70 C4
Rocky River OH 99 D2
Roebuck SC 54 A4
Rogers AR 51 D1
Rogers City MI 16 C4
Rogersville MO 51 E1
Rogersville TN 54 A1
Rogue River OR 17 D2
Rohnert Park CA 31 E3
Roland OK 51 D2
Rolette ND 13 E1
Rolla MO 39 D4
Rolla ND 13 E1
Rolling Fork MS 63 F1
Rolling Hills WY 21 E3
Rolling Meadows IL ... 98 B2
Roma TX 60 C4
Rome GA 53 E3
Rome NY 28 C2
Romeo MI 26 C3
Romeoville IL 98 B6

Philadelphia PA

Entries in **bold color** indicate cities with detailed inset maps.

Romney WV	42 B2	Roselle Park NJ	112 A4	Roundup MT	11 E3	Runnemede NJ	116 D4	Rutherford NJ	112 B2	St. Albans VT	29 D1	St. Clairsville OH	41 F2	Sandusky OH	26 C4

Romney WV 42 B2
Romulus MI 26 C3
Ronan MT 10 A2
Ronceverte WV 41 F4
Roosevelt UT 34 C1
Rosamond CA 44 C3
Rosaryville MD 68 B4
Roscoe IL 25 E3
Roseau MN 14 A1
Rosebud SD 22 C2
Rosebud TX 62 A3
Roseburg OR 17 E2
Rosedale MD 95 D2
Rosedale MS 51 F4
Rose Hill KS 37 F4
Roselawn IN 25 F4
Roselle IL 98 B3
Roselle NJ 112 A4

Roselle Park NJ 112 A4
Rose Lodge OR 17 D1
Rosemead CA 106 D2
Rosemount MN 110 D4
Rosenberg TX 61 E1
Roseville CA 31 F3
Roseville MI 101 H1
Roseville MN 110 C2
Rossville KS 38 A3
Rossville MD 95 D2
Roswell GA 53 F4
Roswell NM 48 B4
Rotan TX 59 D1
Rothschild WI 25 D1
Rotonda FL 67 D3
Rotterdam NY 70 A1
Round Mtn. NV 32 C3
Round Rock TX 62 A3

Roundup MT 11 E3
Rowland Hts. CA 107 F3
Rowlett TX 100 G1
Roxboro NC 55 D1
Roy UT 20 A4
Royal City WA 9 D3
Royal Oak MI 101 H1
Royal Palm Beach FL 67 F3
Royersford PA 68 C2
Royse City TX 62 B1
Royston GA 54 A3
Rubidoux CA 107 J3
Rugby ND 13 E1
Ruidoso NM 48 A4
Ruidoso Downs NM 48 A4
Ruleville MS 52 A4
Rumford ME 29 F1
Rumson NJ 69 F2

Runnemede NJ 116 D4
Rupert ID 19 F3
Rural Hall NC 54 C1
Rush City MN 14 C4
Rushford MN 24 C2
Rush Sprs. OK 50 A3
Rushville IL 39 E2
Rushville IN 40 C2
Rushville NE 22 B3
Ruskin FL 67 D2
Russell KS 37 E3
Russell Sprs. KY 53 E1
Russellville AL 52 C3
Russellville AR 51 E2
Russellville KY 53 D1
Ruston LA 63 D1
Ruth NV 33 E2

S
Sabetha KS 38 A2
Sabina OH 41 D2
Sabinal TX 59 E4
Sac City IA 24 A3
Sacramento CA 31 F3
Saddle Brook NJ 112 B1
Safety Harbor FL 123 D1
Safford AZ 56 C1
Sagamore MA 71 F3
Saginaw MI 26 B2
Sahuarita AZ 56 B2

Rutherford NJ 112 B2
Rutherfordton NC 54 B2
Rutland VT 29 D2

St. Albans VT 29 D1
St. Albans WV 41 E3
St. Ann MO 120 B1
St. Anthony ID 20 A2
St. Augustine FL 66 B4
St. Augustine Beach FL 66 B4
St. Augustine Shores FL 66 B4
St. Charles IL 25 E4
St. Charles MD 42 C3
St. Charles MI 26 B2
St. Charles MN 24 C2
St. Charles MO 39 E3
St. Clair MI 26 C3
St. Clair PA 68 B1
St. Clair Shores MI 101 H1

Ste. Genevieve MO 39 E4
St. Francis KS 36 C2
St. Francis MN 14 C4
St. Francis WI 109 D3
St. Franciville LA 63 F3
St. Gabriel LA 63 F3
St. George SC 66 B1
St. George UT 33 F4
St. Hedwig TX 59 F4
St. Helen MI 26 B1
St. Helena CA 31 E3
St. Helens OR 8 B4
St. Ignace MI 16 C4
St. Ignatius MT 10 A2
St. James MN 24 A2
St. James MO 39 D4
St. John KS 37 E4
St. Johns AZ 47 D3
St. Johns MI 26 B3
St. Johnsbury VT 29 E1
St. Joseph IL 40 A2
St. Joseph MI 25 F4
St. Joseph MN 14 B4
St. Joseph MO 38 B2
St. Louis MI 26 B2
St. Louis MO 39 E3
St. Louis Park MN 110 B2
St. Maries ID 9 F2
St. Martinville LA 63 E4
St. Marys AK 74 B2
St. Marys GA 66 B3
St. Marys KS 38 A2
St. Marys OH 41 D1
St. Marys PA 27 F4
St. Marys WV 41 F2
St. Matthews KY 108 C1
St. Matthews SC 54 B4
St. Michael MN 14 C4
St. Michaels AZ 47 D2
St. Paul AK 74 A3
St. Paul MN 24 B1
St. Paul NE 23 D4
St. Pauls NC 55 D3
St. Pete Beach FL 67 D2
St. Peter MN 24 B1
St. Peters MO 39 E3
St. Petersburg FL 67 D2
St. Robert MO 39 D4
St. Simons Island GA 66 B3
St. Stephen SC 54 C4
St. Thomas ND 13 F1
Salado TX 62 A3
Salamanca NY 27 F3
Salem AR 51 F1
Salem IL 39 F3
Salem IN 40 B3
Salem MA 71 E1
Salem MO 39 D4
Salem NH 71 E1
Salem NJ 68 C3
Salem OH 41 F1
Salem OR 17 E1
Salem SD 23 E2
Salem UT 34 B2
Salem VA 42 A4
Salem WV 41 F2
Salida CA 31 F4
Salida CO 35 F3
Salina KS 37 F3
Salina OK 50 C1
Salina UT 34 A3
Saline MI 26 C3
Salisbury MD 43 D3
Salisbury MA 71 E1
Salisbury MO 38 C2
Salisbury NC 54 C2
Sallisaw OK 50 C2
Salmon ID 10 A4
Salome AZ 46 A4
Saltillo MS 52 B3
Salt Lake City UT 34 A1
Saltville VA 54 B1
Saluda SC 54 B4
Salyersville KY 41 D4
Samson AL 65 D3
Samsula FL 67 E1
San Angelo TX 59 D2
San Augustine TX 62 C2
San Benito TX 61 D4
San Bernardino CA 45 D4
San Bruno CA 31 E4
San Carlos AZ 46 C4
San Carlos CA 122 C5
San Carlos Park FL 67 E3
San Clemente CA 45 D4
San Diego CA 56 B4
San Dimas CA 107 F2

Sandusky OH 26 C4
Sandwich IL 25 E4
Sandwich MA 71 F3
Sandy OR 8 B4
Sandy UT 34 A1
Sandy Sprs. GA 53 F4
Sandy Valley NV 45 E2
San Elizario TX 57 E2
San Felipe Pueblo NM 47 F2
San Fernando CA 44 C3
Sanford FL 67 E1
Sanford ME 29 F2
Sanford NC 55 D2
San Francisco CA 31 E4
San Gabriel CA 106 D2
Sanger CA 44 B3
Sanger TX 50 B4
Sanibel FL 67 D3
San Jacinto CA 45 D4
San Joaquin CA 44 B1
San Jose CA 31 F4
San Juan Capistrano CA 44 C4
San Leandro CA 31 F4
San Lorenzo CA 122 D4
San Luis AZ 56 A2
San Luis Obispo CA 44 A2
San Manuel AZ 56 B1
San Marcos CA 56 B3
San Marcos TX 59 F4
San Martin CA 31 F4
San Mateo CA 31 E4
San Miguel NM 57 E2
San Pablo CA 122 B1
San Rafael CA 31 E4
San Ramon CA 31 F4
San Saba TX 59 F3
Santa Ana CA 44 C4
Santa Barbara CA 44 B3
Santa Clara CA 122 D6
Santa Clara NM 57 D1
Santa Clara OR 17 E1
Santa Clara UT 33 F4
Santa Clarita CA 44 C3
Santa Claus IN 40 B4
Santa Cruz CA 31 F4
Santa Fe NM 47 F2
Santa Fe TX 61 F1
Santa Fe Sprs. CA 106 D3
Santa Maria CA 44 A3
Santa Monica CA 44 C4
Santa Paula CA 44 B3
Santaquin UT 34 A2
Santa Rosa CA 31 E3
Santa Rosa NM 48 B3
Santa Rosa TX 61 D4
Santa Ynez CA 44 B3
Santee CA 56 B4
Santo Domingo Pueblo NM 47 F2
Sapulpa OK 50 B2
Saraland AL 64 B3
Saranac Lake NY 28 C1
Sarasota FL 67 D2
Saratoga CA 31 F4
Saratoga WY 21 D4
Saratoga Sprs. NY 29 D3
Sardis MS 52 A3
Sartell MN 14 B4
Satanta KS 36 C4
Satellite Beach FL 67 F1
Satsuma AL 64 B3
Saucier MS 64 B3
Saugerties NY 70 A2
Sauk Centre MN 14 B4
Sauk City WI 25 D2
Sauk Rapids MN 14 B4
Saukville WI 25 D2
Sault Ste. Marie MI 16 C3
Savage MN 110 B4
Savanna IL 25 D4
Savannah GA 66 B2
Savannah MO 38 B2
Savannah TN 52 C2
Savoonga AK 74 A2
Savoy IL 40 A2
Sawmills NC 54 B2
Sayre OK 49 F3
Sayre PA 28 B4
Sayreville NJ 69 E1
Sayville NY 69 F1
Scappoose OR 8 B4
Schaumburg IL 98 B3
Schenectady NY 70 A1
Schererville IN 25 F4
Schertz TX 121 C1
Schofield WI 25 D1
Schriever LA 63 F4
Schulenburg TX 61 D1
Schurz NV 32 B3
Schuyler NE 23 E4
Schuylkill Haven PA 68 B1
Scituate MA 71 E2
Scobey MT 12 A1
Scotch Plains NJ 70 A3
Scotia NY 70 A1
Scotland SD 23 E2
Scotland Neck NC 55 E2
Scott LA 63 E3
Scott City KS 36 C3
Scott City MO 39 F4
Scottdale GA 94 D2
Scottdale PA 42 A2

Phoenix AZ

Sun City West, Peoria, Sun City, Surprise, Youngtown, El Mirage, Litchfield Park, Goodyear, Avondale, Cashion, Tolleson, Glendale, Phoenix, Paradise Valley, Scottsdale, Fountain Hills, Sunnyslope, Maryvale, Laveen, Guadalupe, Tempe, Mesa, Gilbert, Chandler, Komatke, Gila River Indian Res.

Pittsburgh PA

West View, Fox Chapel, Oakmont, Emsworth, Avalon, Bellevue, McKees Rocks, West Park, Forest Grove, Pittsburgh, Penn Hills, Wilkinsburg, Edgewood, Churchill, Forest Hills, Swissvale, Carnegie, Crafton, Green Tree, Mount Lebanon, Dormont, Castle Shannon, Brentwood, Baldwin, Munhall, North Braddock, Braddock, Turtle Creek, Duquesne, Bridgeville, Bethel Park, Whitehall, West Mifflin, White Oak, McKeesport, Homestead

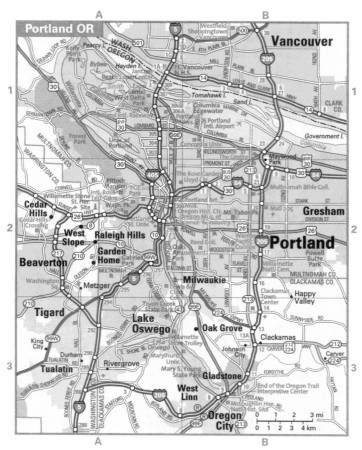

Portland OR

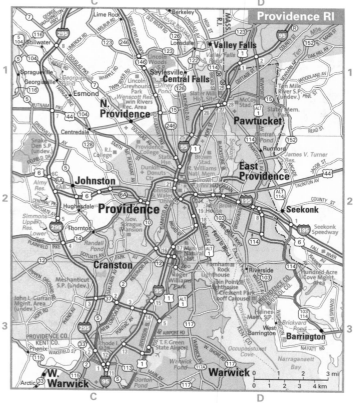

Providence RI

Soldotna AK	74 C3
Soledad CA	44 A1
Scottsburg IN	40 C3
S. Orange NJ	112 A3
S. Padre Island TX	61 D4
S. Paris ME	29 F1
S. Pasadena CA	106 D2
S. Patrick Shores FL	67 F1
S. Pittsburg TN	53 E3
S. Point OH	41 E3
S. Portland ME	29 F2
S. St. Paul MN	110 D3
S. Salt Lake UT	120 C2
S. San Francisco CA	122 B4
S. Sioux City NE	23 F3
Southside AL	53 D4
S. Tucson AZ	56 B2
S. Venice FL	67 D3
S. Whittier CA	106 E3
S. Williamsport PA	28 B4
S. Yarmouth MA	71 F3
Spanaway WA	8 B2
Spanish Fork UT	34 B2
Spanish Fort AL	64 C3
Spanish Lake MO	120 C1
Sparks NV	32 B2
Sparta IL	39 F4
Sparta MI	26 A3
Sparta NJ	43 E1
Sparta TN	53 E2
Sparta WI	25 D2
Spartanburg SC	54 B3
Spearfish SD	21 F1
Spearman TX	49 D2
Speedway IN	40 B2
Spencer IN	40 B3
Spencer IA	24 A2
Spencer NC	54 C2
Spencer WV	41 F3
Spencer WI	25 D1
Spencerport NY	28 A2
Spindale NC	54 B2
Spirit Lake ID	9 E2
Spirit Lake IA	24 A2
Spiro OK	51 D2
Spokane WA	9 E2
Spokane Valley WA	9 E2
Spooner WI	15 D4
Spotswood NJ	69 E2
Spotsylvania VA	42 C3
Spring TX	62 B4
Spring City TN	53 E2
Spring Creek NV	33 E1
Springdale AR	51 D2
Springdale NJ	116 E4
Springdale OH	99 B3
Springer NM	48 B1
Springerville AZ	47 D4
Springfield CO	36 B4
Springfield FL	65 D4
Springfield IL	39 F2
Springfield KY	40 C4
Springfield MA	70 C2
Springfield MN	24 A1
Springfield MO	38 C4
Springfield OH	41 D2
Springfield OR	17 E2
Springfield PA	116 B3
Springfield SD	23 E3
Springfield TN	53 D1
Springfield VT	29 E2
Springhill LA	63 D1
Springhill FL	67 D1
Spring Hill TN	53 D2
Spring Lake MI	26 A3
Spring Lake NJ	69 E2
Spring Lake NC	55 D3
Springs NY	70 C4
Springtown TX	59 F1
Springvale ME	29 F2
Spring Valley CA	56 B4
Spring Valley IL	25 E4
Spring Valley MN	24 C2
Spring Valley NV	105 A2
Spring Valley NY	70 A4
Springville AL	53 D4
Springville UT	34 B1
Spruce Pine NC	54 A2
Stafford KS	37 E4
Stafford TX	102 D3
Stafford VA	42 C3
Stafford Sprs. CT	70 C2
Stamford CT	70 B4
Stamford TX	59 E1
Stamps AR	51 D4
Stanfield OR	9 D4
Stanford KY	40 C4
Stanford MT	11 D2
Stanley NC	54 B2
Stanley ND	12 C1
Stanley WI	25 D1
Stanton CA	106 E4
Stanton NE	23 E4
Stanton TX	58 C2
Stanwood WA	8 B1
Staples MN	14 B3
Star City AR	51 F4
Starke FL	66 A4
Starkville MS	52 B4
State College PA	42 B1
Statesboro GA	66 B1
Statesville NC	54 B2
Statham GA	53 F4
Staunton IL	39 E3
Staunton VA	42 A3
Stayton OR	17 E1
Steamboat Sprs. CO	35 E1
Stearns KY	53 E1
Stebbins AK	74 B2
Steele MO	52 A2
Steele ND	13 E3
Steeleville MO	39 D4
Stephenville TX	59 F1
Sterling CO	36 B3
Sterling IL	25 D4
Sterling KS	37 E3
Sterling VA	68 A4
Sterling Hts. MI	26 C3
Steubenville OH	41 F1
Stevenson AL	53 D3
Stevens Pt. WI	25 D1
Stevensville MD	68 B4
Stevensville MT	10 A3
Stewartville MN	24 B2
Stigler OK	50 C2
Stillwater MN	24 B1
Stillwater OK	50 B2
Stilwell OK	51 D2
Stinnett TX	49 D2
Stockbridge GA	53 F4
Stockton CA	31 F4
Stockton KS	37 E2
Stockton MO	38 C4
Stokesdale NC	54 C1
Stone Mtn. GA	53 F4
Stonewall LA	63 D1
Stonewall MS	64 B2
Stonewood WV	41 F2
Stony Brook NY	69 F1
Stony Pt. NY	70 A4
Storm Lake IA	24 A3
Storrs CT	70 C3
Story WY	21 D1
Story City IA	24 B3
Stoughton MA	71 E2
Stoughton WI	25 E3
Stowell TX	61 F1
Strafford MO	38 C4
Strasburg CO	36 A2
Strasburg ND	13 D3
Strasburg PA	68 B2
Strasburg VA	42 B3
Stratford CT	70 B4
Stratford OK	50 B3
Stratford TX	49 D1
Stratford WI	25 D1
Strawberry AZ	46 B3
Streamwood IL	98 A3
Streator IL	25 E4
Streetsboro OH	27 D4
Stromsburg NE	23 E4
Strongsville OH	27 D4
Stroud OK	50 B2
Stroudsburg PA	43 D1
Stuart FL	67 F2
Stuart IA	24 A4
Stuarts Draft VA	42 A4
Sturgeon Bay WI	25 F1
Sturgis KY	40 A4
Sturgis MI	26 A4
Sturgis SD	22 A1
Sturtevant WI	25 F3

Scottsbluff NE	22 A4
Scottsboro AL	53 D3
Scottsburg IN	40 C3
Scottsdale AZ	46 B4
Scotts Valley CA	31 F4
Scottsville KY	53 D1
Scranton PA	28 B4
Scribner NE	23 F4
Seabrook MD	124 E1
Seabrook NH	71 E1
Seabrook TX	61 F1
Seaford DE	43 D3
Seagoville TX	62 B1
Seagraves TX	58 C1
Seal Beach CA	106 C4
Sealy TX	61 E1
Searchlight NV	45 F2
Searcy AR	51 F2
Searsport ME	30 B3
Seaside CA	31 D4
Seaside OR	8 A4
SeaTac WA	123 B4
Seattle WA	8 B2
Sebastian FL	67 F2
Sebastian TX	61 D4
Sebastopol CA	31 E3
Sebree KY	40 A4
Sebring FL	67 E2
Secaucus NJ	112 C3
Security CO	35 F3
Sedalia MO	38 C3
Sedan KS	38 A4
Sedgwick KS	37 F4
Sedona AZ	46 B3
Sedro-Woolley WA	8 B1
Seeley Lake MT	10 B2
Seguin TX	59 F4
Selah WA	8 C3
Selawik AK	74 B2
Selby SD	13 E4
Selbyville DE	43 D3
Selinsgrove PA	68 B1
Sellersburg IN	40 C3
Sells AZ	56 A2
Selma AL	64 C1
Selma CA	44 B1
Selma NC	55 D2
Selmer TN	52 B3
Seminole OK	50 B2
Seminole TX	58 C1
Senath MO	52 A2
Senatobia MS	52 A3
Seneca KS	38 A2
Seneca MO	51 D1
Seneca SC	54 A3
Seneca Falls NY	28 B3
Sequim WA	8 A2
Sergeant Bluff IA	23 F3
Seth Ward TX	49 D3
Seven Corners VA	124 C1
Seven Hills OH	99 F3
Seven Lakes NC	54 C2
Severn MD	68 B4
Severna Park MD	68 B4
Sevierville TN	53 F2
Seward AK	74 C3
Seward NE	37 F1
Seymour CT	70 B3
Seymour IN	40 B3
Seymour MO	51 E1
Seymour TN	53 F2
Seymour TX	49 F4
Seymour WI	25 E1
Shady Cove OR	17 E3
Shady Side MD	68 B4
Shadyside OH	41 F2
Shady Spr. WV	41 F4
Shafter CA	44 B2
Shaker Hts. OH	99 G2
Shakopee MN	24 B1
Shallowater TX	49 D4
Shamokin PA	68 B1
Shamrock TX	49 E2
Shannon MS	52 B4
Shannon Hills AR	51 E3
Sharon MA	71 E2
Sharon PA	27 E4
Sharon WI	25 E3
Sharonville OH	99 B1
Sharpsburg NC	55 E2
Sharpsville PA	27 E4
Shasta Lake CA	31 E1
Shattuck OK	49 E1
Shaw MS	51 F4
Shawano WI	25 E1
Shawnee KS	38 B3
Shawnee OK	50 B2
Sheboygan WI	25 F2
Sheboygan Falls WI	25 F2
Sheffield AL	52 C3
Shelbina MO	39 D2
Shelburne VT	29 D1
Shelby MS	51 F4
Shelby MT	10 C1
Shelby NC	54 B2
Shelby OH	41 E1
Shelbyville IL	39 F2
Shelbyville IN	40 C2
Shelbyville KY	40 C3
Shelbyville TN	53 D2
Sheldon IA	23 F2
Shelley ID	20 A2
Shelton CT	70 B3
Shelton NE	37 E1
Shelton WA	8 A2
Shenandoah IA	38 B1
Shenandoah VA	42 B3
Shepherd TX	62 C3
Shepherdsville KY	40 C4
Sheridan AR	51 E3
Sheridan IN	40 B2
Sheridan MT	10 C4
Sheridan OR	8 A4
Sheridan WY	21 D1
Sherman IL	39 F2
Sherman TX	50 B4
Sherwood OR	8 B4
Sherwood WI	25 E1
Shields MI	26 B2
Shillington PA	68 C2
Shiner TX	61 D1
Shinnston WV	41 F2
Shippensburg PA	68 A2
Shiprock NM	47 D1
Shirley NY	70 C4
Shishmaref AK	74 B2
Shively KY	108 A2
Shoreline WA	123 B2
Shoreview MN	110 C1
Shorewood IL	25 E4
Shorewood WI	109 D2
Shoshone ID	19 E3
Shoshoni WY	21 D2
Show Low AZ	46 C3
Shreveport LA	63 D1
Shrewsbury MA	71 D2
Shrewsbury PA	68 B3
Sibley IA	23 F2
Sidney MT	12 B2
Sidney NE	22 A4
Sidney NY	28 C3
Sidney OH	41 D1
Sierra Vista AZ	56 B2
Siesta Key FL	67 D2
Signal Mtn. TN	53 E3
Sigourney IA	24 C4
Sikeston MO	52 B1
Siler City NC	54 C2
Siloam Sprs. AR	51 D2
Silsbee TX	62 C3
Silt CO	35 D2
Silver Bay MN	15 D2
Silver City NM	57 D1
Silver Creek NY	27 F3
Silver Hill MD	124 D3
Silver Lake KS	38 A3
Silver Spr. MD	68 A4
Silver Sprs. FL	66 A4
Silver Sprs. NV	32 B2
Silver Sprs. Shores FL	67 E1
Silverthorne CO	35 F2
Silverton NJ	69 E2
Silverton OR	17 E1
Simi Valley CA	44 C3
Simmesport LA	63 E3
Simpsonville SC	54 A3
Simsbury CT	70 C3
Sinclair WY	21 D4
Sinton TX	61 D2
Sioux Ctr. IA	23 F3
Sioux City IA	23 F3
Sioux Falls SD	23 E2
Sisseton SD	13 F4
Sissonville WV	41 E3
Sistersville WV	41 F2
Sitka AK	75 D4
Skagway AK	75 D3
Skaneateles NY	28 B3
Skiatook OK	50 C1
Skidaway Island GA	66 B2
Skidway Lake MI	26 B1
Skippack PA	68 C2
Skokie IL	98 D3
Skowhegan ME	29 F1
Skyway WA	123 C3
Slater MO	38 C2
Slatersville RI	71 D2
Slatington PA	68 C1
Slaton TX	49 D4
Slaughterville OK	50 A3
Slayton MN	23 F2
Sleepy Eye MN	24 A1
Slidell LA	64 A3
Slinger WI	25 E2
Slippery Rock PA	42 A1
Sloatsburg NY	70 A4
Slocomb AL	65 D3
Smackover AR	51 E4
Smith NV	32 B3
Smith Ctr. KS	37 E2
Smithfield NC	55 D2
Smithfield UT	20 A1
Smithfield VA	55 E1
Smiths AL	65 E1
Smithtown NY	69 F1
Smithville MO	38 B2
Smithville TN	53 D2
Smithville TX	62 A4
Smokey Pt.	8 B1
Smyrna DE	68 C4
Smyrna GA	53 E4
Smyrna TN	53 D2
Sneads Ferry NC	55 E3
Snellville GA	53 F4
Snohomish WA	8 B2
Snoqualmie WA	8 B2
Snowflake AZ	46 C3
Snowmass Vil. CO	35 E2
Snyder OK	49 F2
Snyder TX	59 D1
Soap Lake WA	9 D2
Socastee SC	55 D3
Social Circle GA	53 F4
Socorro NM	47 F4
Socorro TX	57 E2
Soda Sprs. ID	20 A3
Soddy-Daisy TN	53 E3
Solana Beach CA	56 A3

Entries in bold color indicate cities with detailed inset maps.
Stuttgart—Waimanalo 119

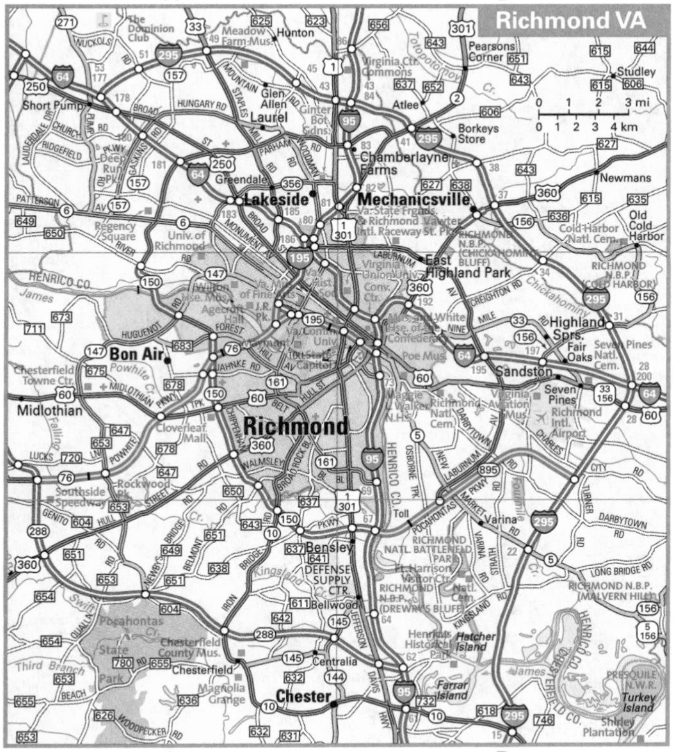
Richmond VA
Richmond
Mechanicsville
Lakeside
Bon Air
Midlothian
Chester
Highland Springs
Sandston
East Highland Park
Bensley
Bellwood
Chesterfield
Centralia
Laurel
Glen Allen
Short Pump
Chamberlayne Farms
Newmans
Studley
Pearsons Corner
Atlee
Borkeys Store
Seven Pines
Varina
Hunton

Stuttgart AR 51 F3
Sublette KS 36 C4
Sublimity OR 17 E1
Suffern NY 70 A4
Suffolk VA 55 F1
Sugar City ID 20 A2
Sugarcreek PA 27 E4
Sugar Hill GA 53 F4
Sugar Land TX 61 E1
Suitland MD 124 E3
Sulligent AL 52 C4
Sullivan IL 39 F2
Sullivan IN 40 A3
Sullivan MO 39 E3
Sullivan City TX 60 C4
Sulphur LA 63 D3
Sulphur OK 50 B3
Sulphur Sprs. TX 62 B1
Sultan WA 8 B2
Sumiton AL 52 C4
Summerfield NC 54 C2
Summersville WV 41 F3
Summerville GA 53 E3
Summerville SC 66 C1
Summit MS 63 F2
Sumner IA 24 C3
Sumner WA 8 B2
Sumter SC 54 C4
Sunburst MT 10 C1
Sunbury OH 41 E1
Sunbury PA 68 B1
Sun City AZ 46 B4
Sun City CA 45 D4
Sun City Ctr. FL 67 D2
Sun City West AZ 117 A1
Suncook NH 29 E2
Sundance WY 21 F1
Sundown TX 48 C4
Sun Lakes AZ 46 B4
Sunland Park NM 57 E2
Sunny Isles Beach FL 109 B3
Sunnyside WA 8 C3
Sunnyvale CA 31 F4
Sun Prairie WI 25 E2
Sunray TX 49 D2
Sunrise FL 109 A2
Sunrise Manor NV 105 B1
Sunset LA 63 E3
Sun Valley ID 19 E2
Superior AZ 46 C4
Superior CO 101 A1
Superior MT 10 A2
Superior NE 37 E2
Superior WI 15 D3
Surfside Beach SC 55 D4
Surprise AZ 46 B4
Surrey ND 13 D1
Susanville CA 32 A1
Sussex WI 25 E2
Sutherland NE 22 B4
Sutherlin OR 17 E2
Sutton AK 74 C3
Sutton NE 37 F1
Suwanee GA 53 F4
Swainsboro GA 66 A1
Swannanoa NC 54 A2
Swanton VT 29 D1
Swartz LA 63 E1
Swartz Creek MI 26 C3
Sweeny TX 61 E1
Sweet Home OR 17 E1
Sweet Sprs. MO 38 C3
Sweetwater FL 67 F4
Sweetwater TN 53 E2
Sweetwater TX 59 D1
Swift Trail Jct. AZ 56 C1
Swissvale PA 117 H2
Switzer WV 41 E4
Sycamore IL 25 E4
Sylacauga AL 65 D1
Sylva NC 53 F2
Sylvania GA 66 B1
Sylvania OH 26 C4
Sylvester GA 65 F2
Syracuse IN 26 A4
Syracuse KS 36 C4
Syracuse NE 38 A1
Syracuse NY 28 B3

T

Tabor City NC 55 D4
Tacoma WA 8 B2
Taft CA 44 B3
Taft TX 61 D2
Tahlequah OK 50 C2
Tahoka TX 58 C1
Takoma Park MD 124 D1
Talent OR 17 E3
Talihina OK 50 C3
Talkeetna AK 74 C3
Talladega AL 53 D4
Tallahassee FL 65 E3
Tallapoosa GA 53 E4
Tallassee AL 65 D1
Talleyville DE 68 C3
Tallulah LA 63 F1
Tama IA 24 B4
Tamaqua PA 68 C1
Tamarac FL 67 F3
Tampa FL 67 D2
Taneytown MD 68 A3
Tanque Verde AZ 56 B1
Taos NM 48 A1
Tappahannock VA 42 C4
Tarboro NC 55 E2
Tarkio MO 38 B1
Tarpon Sprs. FL 67 D1
Tatum NM 48 C4
Taunton MA 71 E2
Tavares FL 67 E1
Tawas City MI 26 C1
Taylor AL 65 D3
Taylor AZ 46 C3
Taylor MI 26 C3
Taylor TX 62 A3
Taylor Creek FL 67 F2
Taylors SC 54 A3
Taylorsville MS 64 A2
Taylorsville UT 120 E3
Taylorville IL 39 F2
Tazewell TN 53 F1
Tazewell VA 54 B1
Tchula MS 52 A4
Tea SD 23 E2
Teague TX 62 B2
Teaneck NJ 112 D1
Tecumseh MI 26 B4
Tecumseh NE 38 A1
Tecumseh OK 50 B2
Tehachapi CA 44 C3
Tekamah NE 23 F4
Tell City IN 40 B4
Telluride CO 35 D4
Temecula CA 45 D4
Tempe AZ 46 B4
Temperance MI 26 C4
Temple GA 53 E4
Temple OK 49 F3
Temple TX 62 A3
Temple City CA 106 E2
Temple Terrace FL 123 E1
Templeton CA 44 A2
Tenafly NJ 112 D1
Tequesta FL 67 F3
Terra Alta WV 42 A2
Terra Bella CA 44 B2
Terrebonne OR 17 F1
Terre Haute IN 40 A2
Terrell TX 62 B1
Terry MT 12 A3
Terrytown LA 111 C2
Tesuque NM 47 F2
Teton ID 20 A2
Tewksbury MA 71 E1
Texarkana AR 51 D4
Texarkana TX 51 D4
Texas City TX 61 F1
Texico NM 48 C3
Thatcher AZ 56 C1
Thayer MO 51 F1
The Colony TX 62 A1
The Dalles OR 8 C4
Theodore AL 64 B3
The Pinery CO 35 F2
Thermalito CA 31 F2
Thermopolis WY 21 D2
The Village OK 50 A2
The Woodlands TX 62 B4
Thibodaux LA 63 F4
Thief River Falls MN 14 A1
Thomas OK 49 F2
Thomaston CT 70 B3
Thomaston GA 65 E1
Thomaston ME 30 B4
Thomasville AL 64 C2
Thomasville GA 65 F3
Thomasville NC 54 C2
Thompson ND 13 F2
Thompson Falls MT 9 F2
Thomson GA 54 A4
Thoreau NM 47 E2
Thorne Bay AK 75 E4
Thornton CO 35 F2
Thorp WI 25 D1
Thorsby AL 64 C1
Thousand Oaks CA 44 C4
Thousand Palms CA 45 D4
Three Forks MT 10 C4
Three Pts. AZ 56 B2
Three Rivers MI 26 A4
Three Rivers TX 60 C2
Thurmont MD 68 A3
Ticonderoga NY 29 D2
Tiffin OH 41 D1
Tifton GA 65 F2
Tigard OR 8 B4
Tilden NE 23 E4
Tillamook OR 8 A4
Tillmans Corner AL 64 B3
Tilton IL 40 A2
Tilton NH 29 E2
Timberville VA 42 B3
Timmonsville SC 54 C4
Tinley Park IL 25 F4
Tinton Falls NJ 69 E2
Tioga ND 12 C1
Tipp City OH 41 D2
Tipton IN 40 B1
Tipton IA 24 C4
Tipton MO 38 C3
Tiptonville TN 52 B1
Tishomingo OK 50 B3
Titusville FL 67 F1
Titusville PA 27 E4
Tiverton RI 71 E3
Toano VA 42 C4
Toccoa GA 53 F3
Togiak AK 74 B3
Tohatchi NM 47 D2
Tok AK 75 D2
Toksook Bay AK 74 A3
Toledo IA 24 B4
Toledo OH 26 C4
Toledo OR 17 D1
Tolleson AZ 46 B4
Tolono IL 40 A2
Tomah WI 25 D2
Tomahawk WI 15 E4
Tomball TX 62 B4
Tombstone AZ 56 C2
Tome NM 47 F3
Tompkinsville KY 53 E1
Toms River NJ 69 E2
Tonawanda NY 97 B2
Tonganoxie KS 38 B3
Tonkawa OK 50 B1
Tonopah NV 32 C4
Tooele UT 34 A1
Tool TX 62 B1
Topeka KS 38 A3
Toppenish WA 8 C3
Topsfield MA 71 E1
Tornillo TX 57 F2
Torrance CA 44 C4
Torrington CT 70 B3
Torrington WY 21 F3
Totowa NJ 112 A1
Towanda KS 37 F4
Towanda PA 28 B4
Towaoc CO 35 D4
Town and Country MO 120 A2
Towner ND 13 D1
Town 'n Country FL 123 D2
Townsend MT 10 C3
Towson MD 68 B3
Tracy CA 31 F4
Tracy MN 23 F1
Tracy City TN 53 D2
Traer IA 24 B3
Travelers Rest SC 54 A3
Traverse City MI 26 A1
Treasure Island FL 67 D2
Tremonton UT 20 A4
Trenton IL 39 F3
Trenton MI 26 C4
Trenton MO 38 C2
Trenton NJ 69 D2
Trenton TN 52 B2
Trent Woods NC 55 E3
Triangle VA 42 C3
Tri-City OR 17 E3
Trinidad CO 36 A4
Trinity AL 52 C3
Trinity TX 62 B3
Tripp SD 23 E2
Triumph LA 64 A4
Troup TX 62 C2
Troy AL 65 D2
Troy ID 9 F3
Troy KS 38 B2
Troy MI 26 C3
Troy MO 39 E3
Troy MT 9 F1
Troy NY 70 B1
Troy NC 54 C2
Troy OH 41 D2
Truckee CA 32 A2
Trumann AR 52 A2
Trumbull CT 70 B4
Trussville AL 53 D4
Truth or Consequences NM 57 E1
Tsaile AZ 47 D1
Tualatin OR 118 A3
Tuba City AZ 46 B2
Tuckerman AR 51 F2
Tuckerton NJ 69 E3
Tucson AZ 56 B1
Tucumcari NM 48 B3
Tukwila WA 123 B4
Tulare CA 44 B1
Tularosa NM 57 F1
Tulia TX 49 D3
Tullahoma TN 53 D2
Tulsa OK 50 C2
Tumwater WA 8 B3
Tunica MS 52 A3
Tupelo MS 52 B3
Tupper Lake NY 28 C1
Turley OK 50 C1
Turlock CA 31 F4
Turners Falls MA 70 C1
Turtle Lake ND 13 D2
Tuscaloosa AL 64 C1
Tuscola IL 40 A2
Tusculum TN 54 A2
Tuscumbia AL 52 C3
Tuskegee AL 65 D1
Tustin CA 107 F5
Tuttle OK 50 A2
Tutwiler MS 52 A4
Twentynine Palms CA 45 E4
Twin Bridges MT 10 B4
Twin Falls ID 19 E3
Twin Lakes NM 47 D2
Twinsburg OH 99 G3
Two Harbors MN 15 D3
Two Rivers WI 25 F1
Tybee Island GA 66 B2
Tyler TX 62 C1
Tylertown MS 64 A3
Tyndall SD 23 E3
Tyrone GA 53 E4
Tyrone PA 42 B1
Tysons Corner VA 124 A2

U

Ucon ID 20 A2
Uhrichsville OH 41 F1
Ukiah CA 31 E2
Ulm MT 10 C2
Ulysses KS 36 C4
Umatilla OR 9 D4
Unadilla GA 65 F1
Unalakleet AK 74 B2
Unalaska AK 74 A4
Underwood ND 13 D2
Unicoi TN 54 A1
Union KY 40 C3
Union MS 64 B1
Union MO 39 E3
Union NJ 69 E1
Union OR 9 E4
Union SC 54 B3
Union City CA 122 D4
Union City GA 53 E4
Union City IN 40 C1
Union City NJ 69 E1
Union City OK 50 A2
Union City PA 27 E4
Union City TN 52 B1
Union Gap WA 8 C3
Union Grove WI 25 F3
Union Sprs. AL 65 D2
Uniontown AL 64 C1
Uniontown PA 42 A2
Unionville CT 70 C3
Unionville MO 38 C1
Unionville NC 54 C3
Universal City TX 59 F4
University City MO 120 B2
University Hts. OH 99 G2
University Park NM 57 E1
University Park TX 100 F2
University Place WA 8 B2
Upland CA 107 G2
Upland IN 40 C1
Upper Arlington OH 100 A2
Upper Sandusky OH 41 D1
Upton WY 21 F1
Urbana IL 40 A2
Urbana OH 41 D2
Urbandale IA 24 B4
Utica NE 37 F1
Utica NY 28 C3
Uvalde TX 59 E4
Uxbridge MA 71 D2

V

Vacaville CA 31 F3
Vadnais Hts. MN 110 D1
Vado NM 57 E2
Vail AZ 56 B2
Vail CO 35 E2
Valdese NC 54 B2
Valdez AK 74 C3
Valdosta GA 65 F3
Vale OR 18 C2
Valencia NM 47 F3
Valentine NE 22 C3
Valier MT 10 C1
Valinda CA 106 E2
Vallejo CA 31 E3
Valley AL 65 E1
Valley NE 23 F4
Valley Ctr. CA 45 D4
Valley Ctr. KS 37 F4
Valley City ND 13 F3
Valley Falls KS 38 A2
Valley Falls RI 118 D1
Valley Sprs. SD 23 F2
Valley Stream NY 113 G4
Valparaiso FL 65 D3
Valparaiso IN 25 F4
Van TX 62 B1
Van Alstyne TX 50 B4
Van Buren AR 51 D2
Van Buren ME 30 C1
Vanceburg KY 41 D3
Vancleave MS 64 B3
Vancouver WA 8 B4
Vandalia IL 39 F3
Vandalia MO 39 D2
Vandalia OH 41 D2
Vandenberg Vil. CA 44 A3
Vandergrift PA 42 A1
Van Horn TX 58 A3
Van Wert OH 40 C1
Varnville SC 66 B1
Vashon WA 123 A4
Vassar MI 26 C2
Vaughn MT 10 C2
Vaughn NM 48 A3
Veedersburg IN 40 A2
Velva ND 13 D2
Veneta OR 17 E2
Venice FL 67 D3
Venice LA 64 A4
Ventnor City NJ 69 E3
Ventura CA 44 B3
Verdi NV 32 A2
Vergennes VT 29 D1
Vermilion OH 26 C4
Vermillion SD 23 E3
Vernal UT 34 C1
Vernon AL 52 C4
Vernon CT 70 C3
Vernon TX 49 F4
Vernon Hills IL 98 C1
Vernonia OR 8 A4
Vero Beach FL 67 F2
Verona MS 52 B4
Verona NJ 112 A2
Verona VA 42 A3
Verona WI 25 D3
Versailles KY 40 C4
Versailles MO 38 C3
Versailles OH 40 C1
Vestal NY 28 B4
Vestavia Hills AL 53 D4
Vian OK 50 C2
Viborg SD 23 E2
Vicksburg MI 26 A3
Vicksburg MS 63 F1
Victor ID 20 B2
Victor MT 10 A3
Victoria KS 37 E3
Victoria MN 24 B1
Victoria TX 61 D2
Victoria VA 55 D1
Victorville CA 45 D3
Vidalia GA 66 A1
Vidalia LA 63 F2
Vidor TX 63 D4
Vienna GA 65 F2
Vienna VA 68 A4
Vienna WV 41 F2
Villa Grove IL 40 A2
Villa Park IL 98 B4
Villa Rica GA 53 E4
Villa Ridge MO 39 E3
Villas NJ 69 D4
Ville Platte LA 63 E3
Vilonia AR 51 E3
Vincennes IN 40 A3
Vincent AL 53 D4
Vincent CA 44 C3
Vine Grove KY 40 B4
Vineland NJ 69 D3
Vinings GA 94 B1
Vinita OK 50 C1
Vinton IA 24 C3
Vinton LA 63 D4
Vinton TX 57 E2
Vinton VA 42 A4
Violet LA 64 A4
Virden IL 39 E2
Virginia MN 14 C2
Virginia Beach VA 55 F1
Virginia City NV 32 B2
Viroqua WI 25 D2
Visalia CA 44 B1
Vista CA 45 D4
Vivian LA 63 D1
Volcano HI 73 F4
Volga SD 23 E1
Voorheesville NY 70 A1

W

Wabash IN 40 B1
Wabasha MN 24 C1
Waco TX 62 A2
Waconia MN 24 B1
Wadena MN 14 B3
Wadesboro NC 54 C3
Wading River NY 70 C4
Wadley GA 66 A1
Wadsworth NV 32 B2
Waggaman LA 111 B2
Wagner SD 23 D3
Wagoner OK 50 C2
Wahiawa HI 72 A2
Wahoo NE 23 F4
Wahpeton ND 14 A3
Waialua HI 72 A2
Waianae HI 72 A3
Waiehu HI 73 D1
Waihee HI 73 D1
Waikoloa Vil. HI 73 E3
Wailua HI 72 B1
Wailuku HI 73 D1
Waimalu HI 72 A3
Waimanalo HI 72 A3

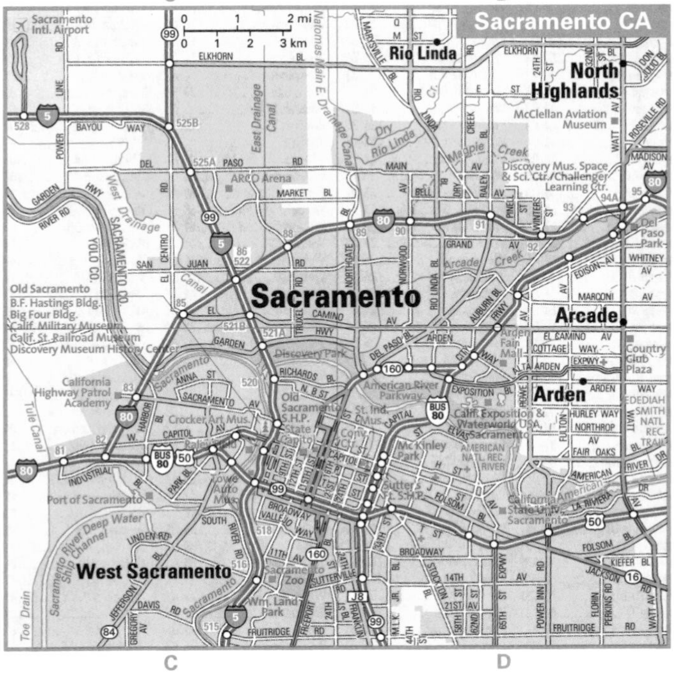
Sacramento CA
Sacramento
West Sacramento
Rio Linda
North Highlands
Arcade
Arden
McClellan Aviation Museum
Sacramento Intl. Airport
ARCO Arena
Old Sacramento
California Highway Patrol Academy
Port of Sacramento

Waimanalo Beach HI 72 A3
Waimea HI 72 B1
Waimea (Kamuela) HI ... 73 E2
Wainwright AK 74 B1
Waipahu HI 72 B1
Waipio Acres HI 72 A3
Waite Park MN 14 B4
WaKeeney KS 37 D3
Wakefield MA 71 E1
Wakefield MI 15 E3
Wakefield NE 23 E3
Wakefield RI 71 D3
Wake Forest NC 55 D2
Wake Vil. TX 51 E4
Walcott IA 25 D4
Walden NY 70 A3
Waldo AR 51 E4
Waldoboro ME 30 B4
Waldport OR 17 D1
Waldron AR 51 D3
Walhalla ND 13 F1
Walhalla SC 54 A1
Walker LA 63 F3
Walker MN 26 A3
Walker Mill MD 124 E2
Walkersville MD 68 A3
Walkerton IN 26 A4
Walkertown NC 54 C2
Wall SD 22 B2
Wallace ID 9 F2
Wallace NC 55 D3
Walla Walla WA 9 E4
Waller TX 62 B4
Wallingford CT 70 C3
Wallingford VT 29 D2
Wallington NJ 112 C2
Walnut CA 107 F3
Walnut Creek CA 31 F4
Walnut Park CA 106 D3
Walnut Ridge AR 51 F2
Walpole MA 71 E2
Walsenburg CO 36 A4
Walterboro SC 66 B1
Walters OK 49 F3
Waltham MA 71 E1
Walthill NE 23 F3
Walthourville GA 66 B2
Walton KY 40 C3
Walton NY 28 C3
Walworth WI 25 E3
Wamego KS 38 A2
Wanaque NJ 70 A4
Wanblee SD 22 B2
Wapakoneta OH 41 D1
Wapato WA 8 C3
Wapello IA 24 C4
Wappingers Falls NY ... 70 A3
Ward AR 51 F3
Warden WA 9 D3
Ware MA 70 C2
Wareham MA 71 E2
Ware Shoals SC 54 A3
Warm Beach WA 8 B1
Warm Sprs. OR 17 F1
Warner OK 50 C2
Warner Robins GA 65 F1
Warr Acres OK 114 E2
Warren AR 51 F4
Warren MI 26 C3
Warren MN 14 A1
Warren OH 27 E4
Warren PA 71 E3
Warrensburg MO 38 C3
Warrensburg NY 29 D2
Warrensville Hts. OH ... 99 G2
Warrenton GA 54 A4
Warrenton MO 39 E3
Warrenton OR 8 A3
Warrenton VA 42 B3
Warrington FL 64 C3
Warrior AL 53 D4
Warroad MN 14 B1
Warsaw IN 26 A4
Warsaw KY 40 C3
Warsaw MO 38 C3
Warsaw NY 27 F3
Warsaw NC 55 D3
Warwick NY 70 A4
Warwick RI 71 D3
Wasco CA 44 B2
Waseca MN 24 B2
Washburn ND 13 D2
Washburn WI 15 E3
Washington DC 68 A4
Washington GA 54 A4
Washington IL 39 F1
Washington IN 40 A3
Washington IA 24 C4
Washington KS 37 F2
Washington MO 39 E3
Washington NJ 69 D1
Washington NC 55 D2
Washington PA 41 F1
Washington UT 33 F4
Washington C.H. OH 41 D2
Washington WA 8 A4
Washingtonville NY 70 A3
Washougal WA 8 B4
Wasilla AK 74 C3
Waskom TX 62 C1
Watauga TX 100 D2
Waterbury CT 70 B3
Waterbury VT 29 D1

Waterford CA 32 A4
Waterford CT 71 D3
Waterford WI 25 E3
Waterloo IL 39 E3
Waterloo IN 26 B4
Waterloo IA 24 C3
Waterloo NY 28 B3
Waterloo WI 25 E2
Watertown CT 70 B3
Watertown MA 96 D1
Watertown MN 24 B1
Watertown NY 28 B2
Watertown SD 23 E1
Watertown WI 25 E2
Water Valley MS 52 B4
Waterville ME 30 B3
Waterville MN 24 B1
Waterville OH 26 C4
Watford City ND 12 B2
Wathena KS 38 B2
Watkinsville GA 53 F4
Watonga OK 49 F2
Watseka IL 40 A1
Watsonville CA 31 D4
Waubay SD 13 F4
Wauchula FL 67 E2
Waukee IA 24 B4
Waukegan IL 25 F3
Waukesha WI 25 E3
Waukomis OK 50 A1
Waukon IA 24 C2
Waunakee WI 25 E2
Waupaca WI 25 E1
Waupun WI 25 E2
Waurika OK 50 A3
Wausau WI 25 D1
Wauseon OH 26 B4
Wautoma WI 25 E2
Wauwatosa WI 109 C2
Waveland MS 64 A4
Waverly IA 24 B3
Waverly NE 37 F1
Waverly NY 28 B4
Waverly OH 41 D3
Waverly TN 52 C2
Waverly VA 55 E1
Waxahachie TX 62 A1
Waxhaw NC 54 B3
Waycross GA 66 A2
Wayland MI 101 E3
Wayne MI 101 E3
Wayne NE 23 E3

Wayne NJ 112 A1
Wayne WV 41 E3
Waynesboro GA 66 A1
Waynesboro MS 64 B2
Waynesboro PA 68 A3
Waynesboro TN 52 C2
Waynesboro VA 42 A3
Waynesburg PA 41 F2
Waynesville MO 39 D4
Waynesville NC 54 A2
Waynesville OH 41 D2
Weatherford OK 49 F2
Weatherford TX 59 F1
Weatherly PA 68 C1
Weaver AL 53 D4
Weaverville CA 31 E1
Weaverville NC 54 A2
Webb City MO 38 B4
Webster MA 71 D2
Webster SD 13 F4
Webster City IA 24 B3
Webster Groves MO ... 120 B3
Weddington NC 54 B3
Weeping Water NE 38 A1
Weigelstown PA 68 B2
Weimar TX 61 D1
Weirton WV 41 F1
Weiser ID 18 C1
Welby CO 101 C2
Welch WV 41 F4
Welcome NC 54 C2
Wellesley MA 71 E2
Wellington CO 35 F1
Wellington FL 67 F3
Wellington KS 37 F4
Wellington OH 27 D4
Wellington TX 49 E1
Wellington UT 34 B2
Wells MN 24 B2
Wells NV 33 E1
Wellsboro PA 28 A4
Wellsburg WV 41 F1
Wellston OH 41 E3
Wellsville KS 38 B3
Wellsville MO 39 D3
Wellsville NY 28 A3
Wellsville OH 41 F1
Wellsville UT 20 A4
Wellton AZ 57 D4
Welsh LA 63 D3
Wenatchee WA 8 C2
Wendell ID 19 E3

Wendell NC 55 D2
Wendover UT 33 F1
Wentworth NC 54 C1
Wentzville MO 39 E3
Weslaco TX 60 C4
Wesley Chapel FL 67 D1
Wesleyville PA 27 F2
Wessington Sprs. SD ... 23 D2
Wesson MS 63 F2
West TX 62 A2
W. Allis WI 25 E3
W. Bend WI 25 E2
Westborough MA 71 D2
W. Branch IA 24 C4
Westbrook ME 29 F2
W. Burlington IA 39 D1
Westby WI 25 D2
Westchester IL 98 C4
W. Chester PA 68 C2
W. Chicago IL 98 A4
W. Columbia TX 61 E1
W. Covina CA 106 E2
W. Crossett AR 63 E1
W. Des Moines IA 24 B4
Westerly RI 71 D3
Westerville OH 41 E2
W. Fargo ND 13 F2
Westfield IN 40 B2
Westfield MA 70 C2
Westfield NJ 27 E3
Westfield TX 62 B4
W. Fork AR 51 D2
W. Frankfort IL 39 F4
W. Grove PA 68 C3
W. Hartford CT 70 C3
W. Haven CT 70 B4
W. Helena AR 52 A3
W. Hollywood CA 106 C2
Westhope ND 13 D1
W. Jefferson OH 41 D2
W. Jordan UT 34 A1
W. Lafayette IN 40 B1
Westlake LA 63 D3
Westlake OH 99 C1
Westland MI 101 E3
W. Liberty IA 24 C4
W. Liberty KY 41 D4
W. Liberty WV 41 F1
W. Linn OR 8 A4
W. Melbourne FL 67 F2
W. Memphis AR 52 A3
Westmere NY 70 A1

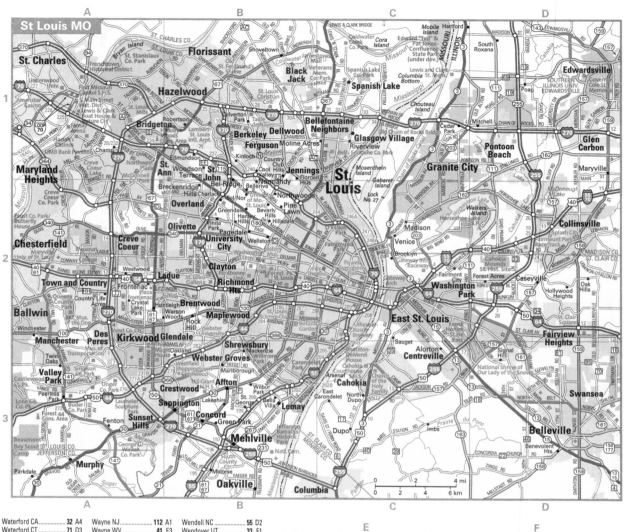

St Louis MO

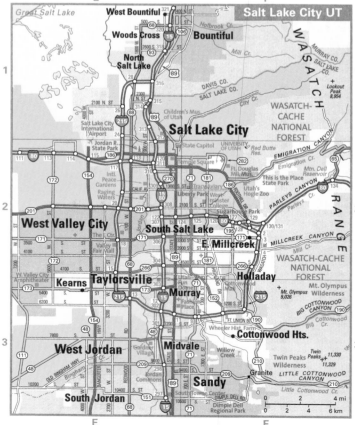

Salt Lake City UT

Entries in **bold color** indicate cities with detailed inset maps.

WEST MIFFLIN—ZWOLLE **121**

W. Mifflin PA 117 G3
W. Milford NJ 70 A4
W. Milton OH 41 D2
Westminster CA 106 E4
Westminster CO 35 F2
Westminster MD 68 A3
Westminster SC 54 A3
W. Monroe LA 63 E1
Westmont IL 98 B5

Westmont PA 42 B1
Westmoreland TN 53 D1
W. New York NJ 112 C2
W. Odessa TX 58 C2
Weston FL 109 A2
Weston MO 38 B2
Weston WV 41 F3
W. Orange NJ 69 E1
W. Orange TX 63 D4

W. Palm Beach FL 67 F3
W. Paterson NJ 112 A1
W. Pensacola FL 64 C3
W. Plains MO 51 F1
W. Point GA 65 E1
W. Point MS 53 B4
W. Point NE 23 F4
W. Point VA 42 C4
Westport CT 70 B4

Westport WA 8 A3
W. Richland WA 9 D3
W. Sacramento CA 31 F3
W. St. Paul MN 110 D3
W. Salem WI 24 C2
W. Seneca NY 27 F3
W. Springfield MA 70 C2
W. Tawakoni TX 62 B1
Westwego LA 64 A4

W. Union OH 41 D3
W. University Place TX 102 D2
W. Valley City UT 34 A1
Westville IL 40 A3
Westville IN 25 F4
Westville OK 51 D2
W. Warwick RI 71 D3

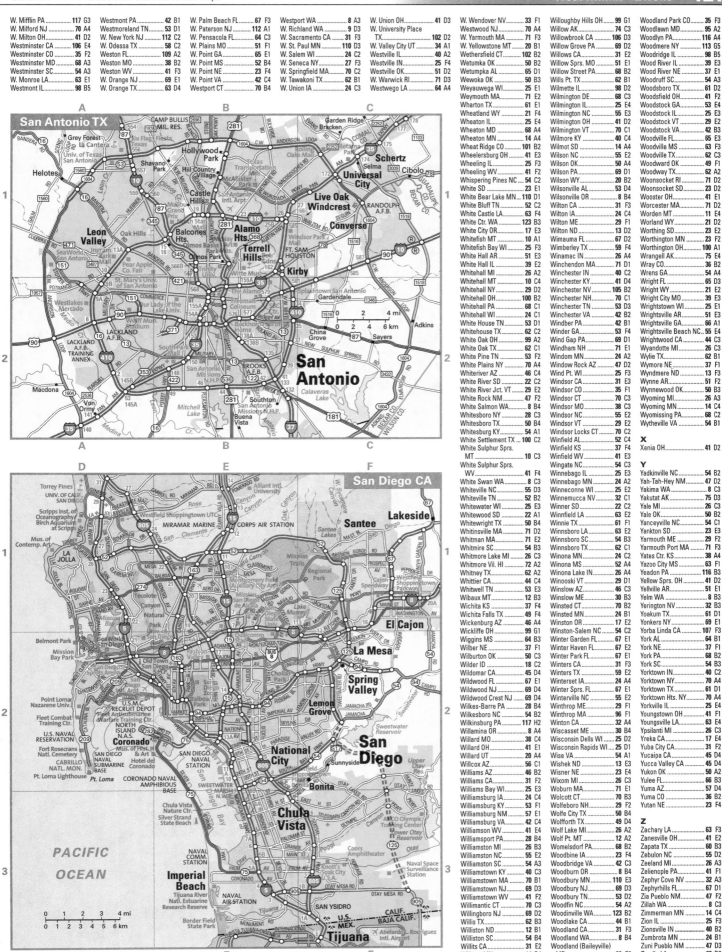

W. Wendover NV 33 F1
Westwood NJ 70 A4
W. Yarmouth MA 71 F3
W. Yellowstone MT 20 B1
Wethersfield CT 102 B2
Wetumka OK 50 B2
Wetumpka AL 65 D1
Wewoka OK 50 B3
Weyauwega WI 25 E1
Weymouth MA 71 E2
Wharton TX 61 E1
Wheatland WY 21 F3
Wheaton IL 25 E4
Wheaton MD 68 A4
Wheaton MN 14 A4
Wheat Ridge CO 101 B2
Wheelersburg OH 41 E3
Wheeling IL 25 F3
Wheeling WV 41 F2
Whispering Pines NC 54 C2
White SD 23 F1
White Bear Lake MN 110 D3
White Castle LA 63 F4
White Ctr. WA 123 B3
White City OR 17 E3
Whitefish MT 10 A1
Whitefish Bay WI 25 F3
White Hall AR 51 E3
White Hall IL 39 E2
Whitehall MI 26 A2
Whitehall MT 10 C4
Whitehall NY 29 D2
Whitehall OH 100 B2
Whitehall PA 68 C1
Whitehall WI 24 C1
White House TN 53 D1
Whitehouse TX 62 C2
White Oak OH 99 A2
White Oak TX 62 C2
White Pine TN 53 F2
White Plains NY 70 A4
Whiteriver AZ 46 C4
White River SD 22 C2
White River Jct. VT 29 E2
White Rock NM 47 F2
White Salmon WA 8 B4
Whitesboro NY 28 C3
Whitesboro TX 50 B4
Whitesburg KY 54 A1
White Settlement TX 100 C2
White Sulphur Sprs.
  MT 10 C3
White Sulphur Sprs.
  WV 41 F4
White Swan WA 8 C3
Whiteville NC 55 D3
Whiteville TN 52 B2
Whitewater WI 25 E2
Whitewood SD 22 A1
Whitewright TX 50 B4
Whitinsville MA 71 D2
Whitman MA 71 E2
Whitmire SC 54 B3
Whitmore Lake MI 26 C3
Whitmore Vil. HI 72 A2
Whitney TX 62 A2
Whittier CA 44 C4
Whitwell TN 53 E3
Wibaux MT 12 B3
Wichita KS 37 F4
Wichita Falls TX 49 F4
Wickenburg AZ 46 A4
Wickliffe OH 99 G1
Wiggins MS 64 B3
Wilber NE 37 F1
Wilburton OK 50 C3
Wilder ID 18 C2
Wildomar CA 45 D4
Wildwood FL 67 E1
Wildwood NJ 69 D4
Wildwood Crest NJ 69 D4
Wilkes-Barre PA 28 B4
Wilkesboro NC 54 B2
Wilkinsburg PA 117 H2
Willamina OR 8 A4
Willard MO 38 C4
Willard OH 41 E1
Willard UT 20 A4
Willcox AZ 56 C1
Williams AZ 46 B2
Williams CA 31 E2
Williams Bay WI 25 E3
Williamsburg IA 24 C4
Williamsburg KY 53 F1
Williamsburg NM 57 E1
Williamsburg VA 42 C4
Williamson WV 41 E4
Williamsport PA 28 B4
Williamston MI 26 B3
Williamston NC 55 E2
Williamston SC 54 A3
Williamstown KY 40 C3
Williamstown MA 70 B1
Williamstown NJ 69 D3
Williamstown WV 41 F2
Willimantic CT 70 C3
Willingboro NJ 69 D2
Willis TX 62 B3
Williston ND 12 B1
Williston SC 54 B4
Willits CA 31 E2
Willmar MN 14 B4

Willoughby Hills OH 99 G1
Willow AK 74 C3
Willowbrook CA 106 D3
Willow Grove PA 69 D2
Willows CA 31 E2
Willow Sprs. MO 51 E1
Willow Street PA 68 B2
Wills Pt. TX 62 B1
Wilmette IL 98 D2
Wilmington DE 68 C3
Wilmington IL 25 E4
Wilmington NC 55 E3
Wilmington OH 41 D2
Wilmington VT 70 C1
Wilmore KY 40 C4
Wilson NC 55 E2
Wilson OK 50 A4
Wilson WY 20 B2
Wilsonville AL 53 D4
Wilsonville OR 8 B4
Wilton CA 31 F3
Wilton IA 24 C4
Wilton ME 29 F1
Wilton ND 13 D2
Wimauma FL 67 D2
Wimberley TX 59 F4
Winamac IN 26 A4
Winchendon MA 71 D1
Winchester IN 40 C2
Winchester KY 41 D4
Winchester NV 105 B2
Winchester NH 70 C1
Winchester TN 53 D3
Winchester VA 42 B2
Windber PA 42 B1
Winder GA 53 F4
Wind Gap PA 69 D1
Windham NH 71 E1
Windom MN 24 A2
Window Rock AZ 47 D2
Wind Pt. WI 25 F3
Windsor CA 31 E3
Windsor CO 35 F1
Windsor CT 70 C3
Windsor MO 38 C3
Windsor NC 55 E2
Windsor VT 29 E2
Windsor Locks CT 70 C3
Winfield AL 52 C4
Winfield KS 37 E3
Winfield WV 41 E3
Wingate NC 54 C3
Winnebago IL 25 E3
Winnebago MN 24 A2
Winneconne WI 25 E1
Winner SD 22 C2
Winnfield LA 63 E2
Winnie TX 61 F1
Winnsboro LA 63 E2
Winnsboro SC 54 B3
Winnsboro TX 62 C1
Winona MN 24 C2
Winona MS 52 A4
Winona Lake IN 26 A4
Winooski VT 29 D1
Winslow AZ 46 C3
Winslow ME 30 B3
Winsted CT 70 B2
Winsted MN 24 B1
Winston OR 17 E2
Winston-Salem NC 54 C2
Winter Garden FL 67 E1
Winter Haven FL 67 E2
Winter Park FL 67 E1
Winters CA 31 F3
Winters TX 59 E2
Winter Sprs. FL 67 E1
Winterville NC 55 E2
Winthrop ME 29 F1
Winthrop MA 96 F1
Winton CA 32 A4
Wiscasset ME 30 B4
Wisconsin Dells WI 25 D2
Wisconsin Rapids WI 25 D1
Wise VA 54 A1
Wishek ND 13 E3
Wisner NE 23 E4
Wixom MI 26 C3
Woburn MA 71 E1
Wolcott CT 70 B3
Wolfeboro NH 29 F2
Wolfe City TX 50 B4
Wolf Lake MI 26 A2
Wolf Pt. MT 12 A2
Womelsdorf PA 68 B2
Woodbine IA 23 F4
Woodbridge VA 42 C3
Woodburn OR 8 B4
Woodbury MN 110 E3
Woodbury NJ 69 D3
Woodbury TN 53 D2
Woodfin NC 54 A2
Woodinville WA 123 B3
Woodlake CA 44 B1
Woodland CA 31 F3
Woodland WA 8 B4
Woodland (Baileyville)
  ME 30 C3

Woodland Park CO 35 F3
Woodlawn MD 95 A2
Woodlyn PA 116 A4
Woodmere NY 113 G5
Woodridge IL 98 B5
Wood River IL 39 E3
Wood River NE 37 E1
Woodruff SC 54 A3
Woodsboro TX 61 D2
Woodsfield OH 41 F2
Woodstock GA 53 E4
Woodstock IL 25 E3
Woodstock VT 29 E2
Woodstock VA 42 B3
Woodville FL 65 E3
Woodville MS 63 F3
Woodville TX 62 C3
Woodward OK 49 F1
Woodway TX 62 A2
Woonsocket RI 71 D2
Woonsocket SD 23 D2
Wooster OH 41 E1
Worcester MA 71 D2
Worden MT 11 E4
Worland WY 21 D2
Worthing SD 23 E2
Worthington MN 23 F2
Worthington OH 100 A1
Wrangell AK 75 E4
Wray CO 36 B2
Wrens GA 54 B4
Wright FL 65 D3
Wright WY 21 E2
Wright City MO 39 E3
Wrightstown WI 25 E1
Wrightsville AR 51 E3
Wrightsville GA 66 A1
Wrightsville Beach NC 55 E4
Wrightwood CA 45 D3
Wyandotte MI 26 C3
Wylie TX 62 B1
Wymore NE 37 F1
Wyndmere ND 13 F3
Wynne AR 51 F2
Wynnewood OK 50 B3
Wyoming MI 26 A2
Wyoming MN 14 C4
Wyomissing PA 68 C2
Wytheville VA 54 B1

X
Xenia OH 41 D2

Y
Yadkinville NC 54 B2
Yah-Tah-Hey NM 47 C3
Yakima WA 8 C3
Yakutat AK 75 D3
Yale MI 26 C3
Yale OK 50 B2
Yanceyville NC 54 C1
Yankton SD 23 E3
Yarmouth ME 29 F2
Yarmouth Port MA 71 F3
Yates Ctr. KS 38 A4
Yazoo City MS 63 F1
Yeadon PA 116 B3
Yellow Sprs. OH 41 D2
Yellville AR 51 E1
Yelm WA 8 B3
Yerington NV 32 B3
Yoakum TX 61 D1
Yonkers NY 69 E1
Yorba Linda CA 107 F3
York AL 64 B1
York NE 37 F1
York PA 68 B2
York SC 54 B3
Yorktown IN 40 C2
Yorktown NY 70 A4
Yorktown TX 61 D1
Yorktown Hts. NY 70 A4
Yorkville IL 25 E4
Youngstown OH 41 F1
Youngsville LA 63 E4
Ypsilanti MI 26 C3
Yreka CA 17 E4
Yuba City CA 31 E2
Yucaipa CA 45 D4
Yucca Valley CA 45 D4
Yukon OK 50 A2
Yulee FL 66 B3
Yuma AZ 57 D4
Yuma CO 36 B2
Yutan NE 23 F4

Z
Zachary LA 63 F3
Zanesville OH 41 E2
Zapata TX 60 B3
Zebulon NC 55 D2
Zeeland MI 26 A3
Zelienople PA 41 F1
Zephyr Cove NV 32 A3
Zephyrhills FL 67 D1
Zia Pueblo NM 47 F2
Zillah WA 8 C3
Zimmerman MN 14 C4
Zion IL 25 F3
Zionsville IN 40 B2
Zumbrota MN 24 B1
Zuni Pueblo NM 47 D3
Zwolle LA 63 E3

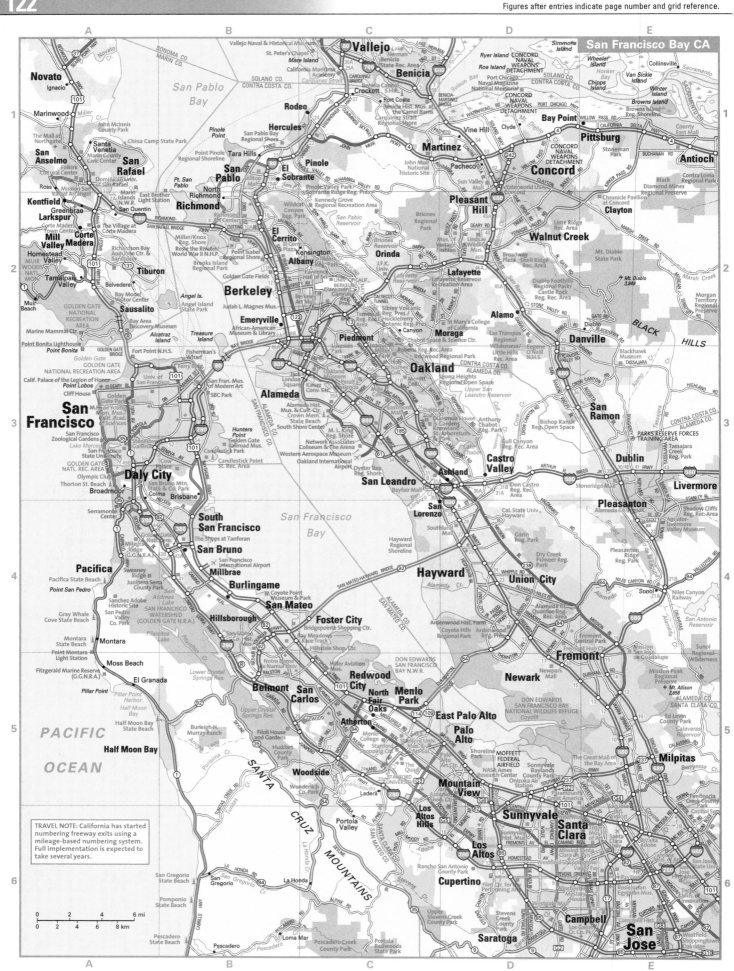

San Francisco Bay CA

TRAVEL NOTE: California has started numbering freeway exits using a mileage-based numbering system. Full implementation is expected to take several years.

Entries in **bold color** indicate cities with detailed inset maps.

Abbotsford—Fermont **123**

CANADA

Abbotsford BC ...78 C4
Acton ON ...84 C3
Acton Vale QC ...87 D4
Airdrie AB ...79 F3
Air Ronge SK ...80 C1
Ajax ON ...84 C3
Alberton PE ...88 C1
Alexandria ON ...85 F1
Alliston ON ...84 C3
Alma QC ...87 D1
Almonte ON ...85 E2
Altona MB ...81 F4
Amherst NS ...88 C3
Amherstburg ON ...84 A4
Amherstview ON ...85 E2
Amos QC ...86 A1
Amqui QC ...89 E4
Angus ON ...84 C2
Antigonish NS ...89 E3
Arborg MB ...81 E2
Armstrong BC ...79 D3
Arnprior ON ...85 E1
Arviat NU ...77 F1
Asbestos QC ...87 D4
Assiniboia SK ...80 C4
Athabasca AB ...79 F1
Atholville NB ...89 E4
Atikokan ON ...82 B3
Attawapiskat ON ...83 E1
Aurora ON ...84 C3
Aylmer ON ...84 B4

Aylmer QC ...85 E1
Ayr ON ...84 B3
Baie-Comeau QC ...87 F1
Baie-Ste-Anne NB ...88 C1
Baie-St-Paul QC ...87 E2
Baie Verte NL ...91 E3
Balgonie SK ...80 C4
Balmoral NB ...89 E4
Bancroft QC ...85 D2
Banff AB ...79 E3
Barrhead AB ...79 E1
Barrie ON ...84 C2
Bas-Caraquet NB ...88 C1
Bathurst NB ...88 C1
Battleford SK ...80 B2
Bay Roberts NL ...91 F4
Beamsville ON ...84 C3
Beauceville QC ...87 E3
Beauharnois QC ...85 F1
Beaumont AB ...79 F1
Beausejour MB ...81 F4
Beauval SK ...80 B1
Beaverlodge AB ...79 D1
Bécancour QC ...87 D3
Belledune NB ...88 B1
Bellefeuille QC ...85 F1
Belleville ON ...85 D3
Beloeil QC ...86 C4
Beresford NB ...88 C1
Berwick NS ...88 C3
Bible Hill NS ...89 D3
Biggar SK ...80 B3

Birch Hills SK ...80 C2
Birtle MB ...81 D4
Bishop's Falls NL ...91 E3
Black Diamond AB ...79 F2
Blackfalds AB ...79 F2
Black Lake QC ...87 D3
Blainville QC ...85 F1
Blairmore AB ...79 F4
Blenheim ON ...84 A4
Blind Bay BC ...79 D3
Blind River ON ...84 A1
Blue Mts. ON ...84 B2
Bluewater ON ...84 B3
Boischatel QC ...87 E3
Boissevain MB ...81 E4
Bolton ON ...84 C3
Bonavista NL ...91 F3
Bonnyville AB ...80 A1
Borden-Carleton PE ...89 D2
Botwood NL ...91 E3
Bouctouche NB ...88 C2
Bow Island AB ...80 A4
Bowmanville ON ...85 D3
Bracebridge ON ...84 C2
Bradford ON ...84 C3
Bradford-W. Gwillimbury ON ...84 C3
Brampton ON ...84 C3
Brandon MB ...81 E4
Brant ON ...84 B3
Brantford ON ...84 C3

Bridgetown NS ...88 C3
Bridgewater NS ...88 C4
Brighton ON ...85 D3
Brockville ON ...85 E2
Bromont QC ...87 D4
Bromptonville QC ...87 D4
Brooklyn NS ...88 C4
Brooks AB ...79 F3
Brownsburg-Chatham QC ...85 E1
Buckingham QC ...85 E1
Buffalo Narrows SK ...80 B1
Burgeo NL ...91 E4
Burin NL ...91 E4
Burlington ON ...84 C3
Burnaby BC ...78 C4
Caledon ON ...84 C3
Caledonia ON ...84 C4
Calgary AB ...79 F3
Calmar AB ...79 F2
Cambridge ON ...84 C3
Campbellford ON ...85 D2
Campbell River BC ...78 B3
Campbellton NB ...89 E4
Camrose AB ...79 F2
Canmore AB ...79 E3
Canora SK ...81 D3
Cantley QC ...85 E1
Cap-de-la-Madeleine QC ...87 D3
Cape Breton R.M. NS ...89 F2

Cap-Pele NB ...88 C2
Capreol ON ...83 E4
Caraquet NB ...88 C1
Carberry MB ...81 E4
Carbonear NL ...91 F4
Cardston AB ...79 F4
Carleton Place ON ...85 E2
Carleton-St-Omer QC ...89 E4
Carlyle SK ...81 D4
Carman MB ...81 E4
Carndoff SK ...81 D4
Caronport SK ...80 C4
Carrot River SK ...80 C2
Carstairs AB ...79 F2
Castlegar BC ...79 E4
Cedar BC ...78 B4
Centreville NB ...88 C2
Chambly QC ...86 C4
Channel-Port aux Basques NL ...91 D4
Chapleau ON ...83 E4
Charlo NB ...89 E4
Charlottetown PE ...89 D2
Charny QC ...87 E3
Chase BC ...79 D3
Chatham ON ...84 A4
Chatham-Kent ON ...84 A4
Chertsey QC ...85 E1
Chester NS ...88 C3
Chestermere AB ...79 F3

Chetwynd BC ...75 F4
Chibougamau QC ...87 F4
Chicoutimi (Saguenay) QC ...87 D1
Chilliwack BC ...78 C4
Chipman NB ...88 C1
Churchill MB ...77 F2
Clarence-Rockland ON ...85 F1
Clarenville NL ...91 F3
Claresholm AB ...79 F3
Clarington ON ...85 D3
Coaldale AB ...79 F3
Coaticook QC ...87 D4
Cobalt ON ...83 F4
Cobourg ON ...85 D3
Cocagne NB ...88 C2
Cochrane AB ...79 F3
Cochrane ON ...83 E4
Coldbrook NS ...88 C3
Cold Lake AB ...80 A1
Coldstream BC ...79 D3
Collingwood ON ...84 C2
Comox BC ...78 B4
Conception Bay S. NL ...91 F4
Contrecoeur QC ...86 C4
Coquitlam BC ...78 C4
Corner Brook NL ...91 D3
Cornwall ON ...85 F2
Cornwall PE ...89 D2
Coronach SK ...80 C4
Coteau-du-Lac QC ...85 F1
Courtenay BC ...78 B4

Cowansville QC ...87 D4
Cranbrook BC ...79 E4
Creighton SK ...81 D1
Creston BC ...79 E4
Crossfield AB ...79 F3
Crowsnest Pass AB ...79 F4
Crystal Beach ON ...84 C4
Dalhousie NB ...89 E4
Dalmeny SK ...80 B3
Danville QC ...87 D4
Dauphin MB ...81 E3
Davidson SK ...80 C3
Dawson YT ...75 D2
Dawson Creek BC ...75 F4
Deep River ON ...85 D1
Deer Lake NL ...91 D3
Dégelis QC ...87 F2
Delhi ON ...84 B4
Delisle QC ...87 D1
Delisle SK ...80 B3
Deloraine MB ...81 E4
Delta BC ...78 C4
Deschambault Lake SK ...81 D1
Des Ruisseaux QC ...86 B3
Devon AB ...79 F1
Didsbury AB ...79 F2
Dieppe NB ...88 C2
Digby NS ...88 B3
Dolbeau-Mistassini QC ...87 D1
Donnacona QC ...87 D3
Drayton Valley AB ...79 E2
Drumheller AB ...79 F2

Drummondville QC ...87 D4
Dryden ON ...82 A2
Duncan BC ...78 B4
Dunnville ON ...84 C4
Dupuy QC ...87 F4
E. Angus QC ...87 D4
E. Gwillimbury ON ...84 C3
Edmonton AB ...79 F1
Edmundston NB ...87 F2
Edson AB ...79 E1
Eel River Crossing NB ...89 E4
Elkford BC ...79 E3
Elliot Lake ON ...84 A1
Elmira ON ...84 B3
Elmsdale NS ...89 D3
Embrun ON ...85 F2
Enderby BC ...79 D3
Enfield NS ...89 D3
Englehart ON ...83 F4
Erin ON ...84 C3
Espanola ON ...84 A1
Essex ON ...84 A4
Esterhazy SK ...81 D3
Estevan SK ...81 D4
Eston SK ...80 B3
Exeter ON ...84 B3
Fairview AB ...76 A3
Falher AB ...76 A3
Fall River NS ...89 D3
Farnham QC ...87 D4
Fergus ON ...84 C3
Fermont QC ...90 A1

Seattle/Tacoma WA

Tampa/St Petersburg FL

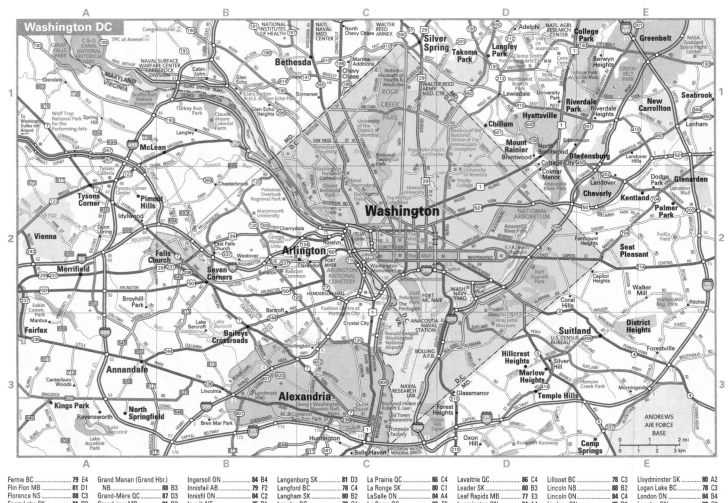

Washington DC

Fernie BC	79 E4	Grand Manan (Grand Hbr.)		Ingersoll ON	84 B4	Langenburg SK	81 D3	La Prairie QC	86 C4	Lavaltrie QC	86 C4	Lillooet BC	78 C3	Lloydminster SK	80 A2
Flin Flon MB	81 D1	NB	88 B3	Innisfail AB	79 F2	Langford SK	78 C4	La Ronge SK	80 C1	Leader SK	80 B3	Lincoln NB	88 B2	Logan Lake BC	78 C3
Florence NS	88 C3	Grand-Mère QC	87 D3	Innisfil ON	84 C2	Langham SK	80 B3	LaSalle ON	84 A4	Leaf Rapids MB	77 E3	Lincoln ON	84 C4	London ON	84 B4
Foam Lake SK	81 D3	Grandview MB	81 D3	Inuvik NT	75 D1	Langley BC	78 C4	La Sarre QC	83 F3	Leamington ON	84 A4	Lindsay ON	85 D2	Longlac ON	82 C3
Forestville QC	87 F1	Gravelbourg SK	80 B4	Invermere BC	79 E3	Lanigan SK	80 C3	La Scie NL	91 E2	Leduc AB	79 F2	L'Islet QC	87 E2	Longueuil QC	86 C4
Ft. Chipewyan AB	76 B2	Gravenhurst ON	84 C2	Inverness NS	89 E2	Lantz NS	89 D3	L'Assomption QC	86 C4	Lennoxville QC	87 D4	Listowel ON	84 B3	Lorraine QC	83 F4
Ft. Erie ON	84 C4	Greater Napanee ON	85 E2	Iqaluit NU	7	La Pêche QC	85 E1	Laterrière QC	87 E1	Lively ON	83 E4	Louisbourg NS	89 F2
Ft. Frances ON	82 A3	Greater Sudbury ON	84 B1	Irishtown-Summerside		La Pocatière QC	87 E2	La Tuque QC	87 D2	Lévis QC	87 E3	Liverpool NS	88 C4	Louiseville QC	87 D3
Ft. Macleod AB	79 F4	Greenstone ON	82 C3	NL	91 D3	Lappe ON	82 B3	Laval QC	85 F3	Lewisporte NL	91 E3	Lloydminster AB	80 A2	Lumsden SK	80 C3
Ft. McMurray AB	76 B3	Greenwood NS	88 C3	Iroquois Falls ON	83 E3						
Ft. Nelson BC	75 F3	Grenfell SK	81 D4	Jasper AB	79 D2						
Ft. Qu'Appelle SK	80 C3	Grimsby ON	84 C3	Joliette QC	86 C3						
Ft. St. John BC	75 F4	Grimshaw AB	76 A3	Jonquière QC	87 D1						
Ft. Saskatchewan AB	79 F1	Guelph ON	84 C3	Kamloops BC	79 D3						
Ft. Simpson NT	76 A1	Gull Lake SK	80 B4	Kamsack SK	81 D3						
Ft. Smith NT	76 B2	Haileybury ON	83 F4	Kanata ON	85 E1						
Fortune NL	91 E4	Haines Jct. YT	75 D3	Kapuskasing ON	83 E3						
Ft. Vermilion AB	76 A2	Haldimand ON	84 C4	Kawartha Lakes ON	84 C2						
Fox Creek AB	79 E1	Halifax NS	89 D3	Kelowna BC	79 D4						
Fredericton NB	88 B2	Halton Hills ON	84 C3	Kelvington SK	81 D3						
French Creek BC	78 B4	Hamilton ON	84 C3	Kemptville ON	85 E2						
Fruitvale BC	79 E4	Hamiota MB	81 E4	Kenora ON	82 A2						
Gabriola Island BC	78 B4	Hammonds Plains Road		Kensington PE	89 D2						
Gambo NL	91 F3	NS	89 D3	Kentville NS	88 C3						
Gananoque ON	85 E2	Hampton NB	88 B3	Kerrobert SK	80 B3						
Gander NL	91 E3	Hanna AB	79 F2	Keswick Ridge NB	88 B2						
Gaspé QC	89 F4	Hanover ON	84 C2	Kimberley BC	79 E4						
Gatineau QC	85 E1	Hantsport NS	88 C3	Kincardine ON	84 B3						
Georgetown ON	84 C3	Happy Valley-Goose Bay		Kindersley SK	80 B3						
Georgetown PE	89 D2	NL	90 C1	Kingston NS	88 C3						
Georgina ON	84 C2	Harbour Breton NL	91 E4	Kingston ON	85 E2						
Geraldton ON	82 C3	Harbour Grace NL	91 F3	Kingsville ON	84 A4						
Gibbons AB	79 F1	Hare Bay NL	91 F3	Kipling SK	81 D4						
Gift Lake AB	76 A3	Havre-St-Pierre QC	90 B2	Kippens NL	91 D3						
Gilbert Plains MB	81 E3	Hawkesbury ON	85 F1	Kirkland Lake ON	83 F3						
Gillam MB	77 F3	Hay River NT	76 B1	Kitchener ON	84 B3						
Gimli MB	81 F3	Hearst ON	83 E3	Kitimat BC	78 A1						
Glace Bay NS	89 F2	Herbert SK	80 B4	Kugluktuk NU	75 F1						
Gladstone MB	81 E4	Herring Cove NS	89 D3	La Baie QC	87 E1						
Glovertown NL	91 F3	High Level AB	76 A2	Labrador City NL	90 A1						
Goderich ON	84 B3	High Prairie AB	76 A3	Lac-Brome QC	87 D4						
Golden BC	79 E3	High River AB	79 F2	Lac du Bonnet MB	81 F4						
Gold River NS	88 C3	Hinton AB	79 E2	Lachute QC	85 F1						
Granby QC	87 D4	Holyrood NL	91 F4	Lac La Biche AB	76 B3						
Grand Bank NL	91 E4	Hope BC	78 C4	Lac-Mégantic QC	87 E4						
Grand Bay-Westfield		Hornepayne ON	83 D3	Lacombe AB	79 F2						
NB	88 B3	Houston BC	78 B1	La Crête AB	76 A2						
Grand Ctr. AB	80 A1	Hudson Bay SK	81 D2	Ladysmith BC	78 B4						
Grande Cache AB	79 D1	Hudson's Hope BC	75 F4	Lake Country BC	79 D3						
Grande-Digue NB	88 C2	Hull QC	85 E1	Lake Echo NS	89 D3						
Grande Prairie AB	79 D1	Humber Arm South NL	91 D3	Lakeshore ON	84 A4						
Grande-Rivière QC	89 F4	Humboldt SK	80 C3	La Loche SK	76 C3						
Grand Falls (Grand Sault)		Huntsville ON	84 C2	La Malbaie QC	87 E2						
NB	88 A1	Huron East ON	84 B3	Lambton Shores ON	84 B3						
Grand Falls-Windsor		Iberville QC	87 D4	Lamèque NB	88 C1						
NL	91 E3	Île-à-la-Crosse SK	80 B1	Lamont AB	79 F1						
Grand Forks BC	79 D4	Indian Head SK	80 C4						

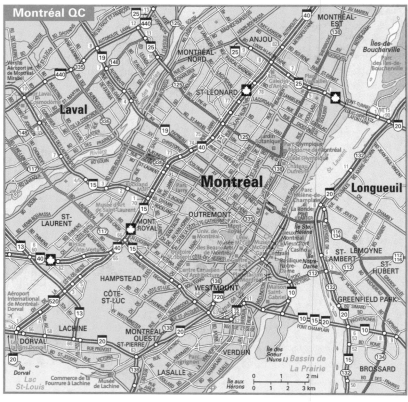

Montréal QC

Entries in **bold color** indicate cities with detailed inset maps.
Lunenburg—Yorkton 125

Toronto ON

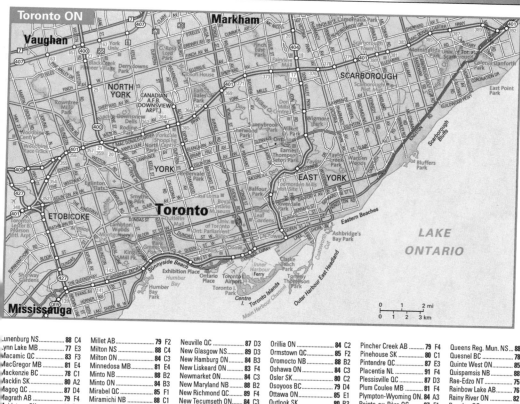

unenburg NS 88 C4
ynn Lake MB 77 E3
Macamic QC 83 F3
MacGregor MB 81 E4
Mackenzie BC 78 C1
Macklin SK 80 A2
Magog QC 87 D4
Magrath AB 79 F4
Maidstone SK 80 B2
Malartic QC 86 A1
Manitou MB 81 E4
Manitouwadge ON 83 D3
Maniwaki QC 85 E1
Manning AB 76 A3
Maple Creek SK 80 A4
arathon ON 82 C3
arieville QC 86 C4
arkham ON 84 C3
artensville SK 80 C3
arystown NL 91 E4
ascouche QC 86 C4
asson-Angers QC 85 E1
atagami QC 83 F2
atane QC 87 F1
ayerthorpe AB 79 E1
cAdam NB 88 B3
cLennan AB 76 A3
cWatters QC 83 F3
eadow Lake SK 80 B1
eaford ON 84 B2
dicine Hat AB 80 A4
lfort SK 80 C2
lita MB 81 D4
lville SK 81 D3
mramcook NB 88 C2
rritt BC 78 C3
tabetchouan-Lac-à-la-roix QC 87 D1
dleton NS 88 C3
land ON 84 C2
Millet AB 79 F2
Milton NS 88 C4
Milton ON 84 C3
Minnedosa MB 81 E4
Minto NB 88 B2
Minto ON 84 B3
Mirabel QC 85 F1
Miramichi NB 88 C1
Mira Road NS 89 F2
Miscouche PE 89 D2
Mission BC 78 C4
Mississauga ON 84 C3
Mississippi Mills ON 85 E2
Mitchell ON 84 B3
Moncton NB 88 C2
Mono ON 84 C3
Montague PE 89 D2
Mont-Joli QC 87 F1
Mont-Laurier QC 86 B3
Montmagny QC 87 E3
Montréal QC 86 C4
Moose Factory ON 83 E2
Moose Jaw SK 80 C4
Moose Lake MB 81 D2
Moosomin SK 81 D4
Moosonee ON 83 E2
Morden MB 81 E4
Morinville AB 79 F1
Morris MB 81 F4
Mt. Forest ON 84 B3
Mt. Pearl NL 91 F4
Musgrave Hbr. NL 91 F3
Nanaimo BC 78 B4
Nanoose Bay BC 78 B4
Nanton AB 79 F3
Napanee ON 85 E2
Nauwigewauk NB 88 B3
Neepawa MB 81 E4
Neguac NB 88 C1
Nelson BC 79 E4
Neuville QC 87 D3
New Glasgow NS 89 D3
New Hamburg ON 84 B3
New Liskeard ON 83 F4
Newmarket ON 84 C3
New Maryland NB 88 B2
New Richmond QC 89 F4
New Tecumseth ON 84 C3
New Waterford NS 89 F2
New Westminster BC 125 C2
New-Wes-Valley NL 91 F3
Niagara Falls ON 84 C4
Niagara-on-the-Lake ON 84 C3
Nicolet QC 87 D3
Nipawin SK 80 C2
Nipigon ON 82 C3
Niverville MB 81 F4
Norfolk ON 84 B4
Normandin QC 87 D1
Normétal QC 83 F3
N. Battleford SK 80 B2
N. Bay ON 84 C1
N. Perth ON 84 B3
N. Vancouver BC 125 B1
N. Vancouver (DM) BC 125 C1
Norton NB 88 B3
Notre-Dame-du-Mont-Carmel QC 87 D3
Notre-Dame-du-Nord QC 83 F4
Oak Bay BC 78 C4
Oakville ON 84 C3
Okotoks AB 79 F3
Olds AB 79 F2
O'Leary PE 88 C2
Oliver BC 79 D4
Onaping ON 83 E4
108 Mile Ranch BC 78 C3
Orangeville ON 84 C3
Orillia ON 84 C2
Ormstown QC 85 F2
Oromocto NB 88 B2
Oshawa ON 84 C3
Osler SK 80 C2
Osoyoos BC 79 D4
Ottawa ON 85 E1
Outlook SK 80 B3
Owen Sound ON 84 B2
Oxbow SK 81 D4
Oxford NS 89 D2
Palmarolle QC 83 F3
Paradise NL 91 F4
Paris ON 84 B3
Parksville BC 78 B4
Parrsboro NS 88 C3
Parry Sound ON 84 C1
Pasadena NL 91 D3
Paspébiac QC 89 F4
Peace River AB 76 A3
Peachland BC 79 D4
Pelham ON 84 C4
Pembroke ON 85 D1
Penetanguishene ON 84 C2
Penhold AB 79 F2
Penticton BC 79 D4
Percé QC 90 B3
Perth ON 85 E2
Perth-Andover NB 88 A1
Petawawa ON 85 D1
Peterborough ON 85 D2
Petitcodiac NB 88 C2
Petit-Rocher NB 88 C1
Petrolia ON 84 A4
Pickering ON 84 C3
Picton ON 85 D3
Pictou NS 89 D2
Picture Butte AB 79 F3
Pilot Butte SK 80 C4
Pinawa MB 81 F4
Pincher Creek AB 79 F4
Pinehouse SK 80 C1
Pintendre QC 87 E3
Placentia NL 91 F4
Plessisville QC 87 D3
Plum Coulee MB 81 F4
Plympton-Wyoming ON 84 A3
Pointe-au-Père QC 87 F1
Ponoka AB 79 F2
Pontiac QC 85 E1
Pont-Landry NB 88 C1
Pont-Rouge QC 87 D3
Porcupine ON 83 E3
Porcupine Plain SK 81 D2
Portage la Prairie MB 81 E4
Port Alberni BC 78 B4
Port-Cartier QC 90 A2
Port Colborne ON 84 C4
Port Dover ON 84 C4
Port Elgin ON 84 B2
Port Hardy BC 78 A3
Port Hawkesbury NS 89 E3
Port Hope ON 85 D3
Port McNeill BC 78 A3
Port Perry ON 84 C3
Pouch Cove NL 91 F4
Powell River BC 78 B4
Powerview MB 81 F3
Preeceville SK 81 D3
Prescott ON 85 E2
Prévost QC 85 F1
Prince Albert SK 80 C2
Prince Edward ON 85 D3
Prince George BC 78 C1
Prince Rupert BC 78 A1
Princeton BC 79 D4
Princeville QC 87 D3
Provost AB 80 A2
Qualicum Beach BC 78 B4
Québec QC 87 E3
Queens Reg. Mun. NS 88 C4
Quesnel BC 78 C2
Quinte West ON 85 D3
Quispamsis NB 88 B3
Rae-Edzo NT 75 F2
Rainbow Lake AB 76 A2
Rainy River ON 82 A3
Rawdon QC 85 F1
Raymond AB 79 F4
Redcliff AB 80 A4
Red Deer AB 79 F2
Red Lake ON 82 A2
Redvers SK 81 D4
Redwater AB 79 F1
Regina SK 80 C4
Regina Beach SK 80 C3
Renfrew ON 85 E1
Repentigny QC 86 C4
Reserve Mines NS 89 F2
Revelstoke BC 79 D3
Richibucto NB 88 C2
Richmond BC 78 C4
Richmond QC 87 D4
Richmond Hill ON 84 C3
Ridgetown ON 84 B4
Rigaud QC 85 F1
Rimbey AB 79 F2
Rimouski QC 87 F1
Rivers MB 81 E4
Riverview NB 88 C2
Rivière-du-Loup QC 87 E2
Roberval QC 87 D1
Roblin MB 81 D3
Rocanville SK 81 D4
Rock Forest QC 87 D4
Rockland ON 85 F1
Rocky Mtn. House AB 79 E2
Rosetown SK 80 B3
Rossland BC 79 E4
Rosthern SK 80 C2
Rothesay NB 88 B3
Rouyn-Noranda QC 83 F3
Roxton Pond QC 87 D4
Rusagonis NB 88 B2
Russell MB 81 D3
Russell ON 85 F2
Saanich BC 78 C4
Sackville NB 88 C2
St. Adolphe MB 81 F4
St. Alban's NL 91 E4
St. Albert AB 79 F1
St-Ambroise QC 87 D1
St. Andrews NB 88 B3
St. Anthony NL 91 E2
St-Antoine NB 88 C2
St-Antonin QC 87 E2
St-Apollinaire QC 87 D3
St-Boniface-de-Shawinigan QC 87 D3
St-Bruno-de-Guigues QC 83 F4
St. Catharines ON 84 C3
St-Césaire QC 87 D4
St-Charles-Borromée QC 86 C3
St-Cyrille-de-Wendover QC 87 D4
Ste-Adèle QC 85 F1
Ste-Agathe-des-Monts QC 85 F1
Ste. Anne MB 81 F4
Ste-Anne-des-Monts QC 89 E3
Ste-Julie QC 86 C4
Ste-Marie QC 87 E3
Ste-Rosalie QC 87 D4
Ste. Rose du Lac MB 81 E3
St-Eustache QC 85 F1
St-Félicien QC 87 D1
St-Félix-de-Valois QC 86 C3
St. George NB 88 B3
St. George's NL 91 D3
St-Georges QC 87 D3
St-Georges QC 87 E3
St-Henri QC 87 E3
St-Honoré QC 87 E1
St-Hyacinthe QC 87 D4
St-Jean-de-Matha QC 86 C3
St-Jean-Port-Joli QC 87 E2
St-Jean-sur-Richelieu QC 86 C4
St-Jérôme QC 85 F1
Saint John NB 88 B3
St. John's NL 91 F4
St-Joseph-de-Beauce QC 87 E3
St. Lawrence NL 91 E4
St. Leonard NB 87 F2
St-Lin-Laurentides QC 85 F1
St. Marys ON 84 B3
St-Nicéphore QC 87 D4
St-Nicolas QC 87 D3
St-Pascal QC 87 E2
St. Paul AB 79 F1
St-Pierre-Jolys MB 81 F4
St-Prosper QC 87 E3
St-Quentin NB 88 B1
St-Raymond QC 87 D3
St. Stephen NB 88 B3
St. Thomas ON 84 B4
St-Timothée QC 85 F1
St-Tite QC 87 D3
St-Zotique QC 85 F1
Salaberry-de-Valleyfield QC 85 F1
Salisbury NB 88 C2
Salmon Arm BC 79 D3
Salmon River NS 88 B4
Sandy Bay SK 81 D1
Sarnia ON 84 A4
Saskatoon SK 80 C3
Saugeen Shores ON 84 B2
Sault Ste. Marie ON 83 D4
Sechelt BC 78 B4
Selkirk MB 81 F4
Sept-Îles QC 90 A2
Sexsmith AB 76 A3
Shannon QC 87 D3
Shaunavon SK 80 B4
Shawinigan QC 87 D3
Shediac NB 88 C2
Shelburne NS 88 C4
Shellbrook SK 80 C2
Sherbrooke QC 87 D4
Shippagan NB 88 C1
Shoal Lake MB 81 D4
Sicamous BC 79 D3
Sidney BC 78 C4
Simcoe ON 84 C4
Sioux Lookout ON 82 B2
Slave Lake AB 76 A4
Smithers BC 78 B1
Smiths Falls ON 85 E2
Smithville ON 84 C4
Smooth Rock Falls ON 83 E3
Snow Lake MB 81 E1
Sooke BC 78 B4
Sorel-Tracy QC 86 C3
Souris MB 81 E4
Souris PE 89 D2
Southampton ON 84 B2
S. Bruce Peninsula ON 84 B2
Southend SK 77 D3
S. Huron ON 84 B3
S. Indian Lake MB 77 E3
Spallumcheen BC 79 D3
Spaniard's Bay NL 91 F4
Sparwood BC 79 F4
Spirit River AB 76 A3
Spiritwood SK 80 B2
Springdale NL 91 E3
Springhill NS 88 C3
Spruce Grove AB 79 F1
Squamish BC 78 C4
Stayner ON 84 C2
Steinbach MB 81 F4
Stellarton NS 89 D3
Stephenville NL 91 D3
Stephenville Crossing NL 91 D3
Stettler AB 79 F2
Stewiacke NS 89 D3
Stonewall MB 81 F4
Stony Mtn. MB 81 F4
Stony Plain AB 79 F1
Stouffville ON 84 C3
Stratford ON 84 B3
Stratford PE 89 D2
Strathmore AB 79 F3
Strathroy ON 84 B4
Sturgeon Falls ON 84 C1
Sudbury ON 83 E4
Summerland BC 79 D4
Summerside PE 89 D2
Sundre AB 79 F2
Surrey BC 78 C4
Sussex NB 88 C2
Sussex Corner NB 88 C2
Sutton ON 84 C2
Swan Hills AB 79 E1
Swan River MB 81 D3
Swift Current SK 80 B4
Sydney NS 89 F2
Sydney Mines NS 89 F2
Sydney River NS 89 F2
Sylvan Lake AB 79 F2
Taber AB 79 F3
Tecumseh ON 84 A4
Temagami ON 83 F4
Témiscaming QC 83 F4
Terrace BC 78 A1
Terrace Bay ON 82 C3
Terrebonne QC 85 F1
Teulon MB 81 F4
The Pas MB 81 D2
Thetford Mines QC 87 D3
Thompson MB 81 E1
Thorold ON 84 C3
Three Hills AB 79 F2
Thunder Bay ON 82 B3
Tignish PE 88 C1
Tilbury ON 84 A4
Tillsonburg ON 84 B4
Timberlea NS 89 D3
Timmins ON 83 E3
Tisdale SK 80 C2
Tofield AB 79 F1
Torbay NL 91 F4
Toronto ON 84 C3
Tracadie-Sheila NB 88 C1
Trail BC 79 E4
Trent Hills ON 85 D3
Trenton NS 89 D3
Triton NL 91 E3
Trois-Pistoles QC 87 F2
Trois-Rivières QC 87 D3
Truro NS 89 D3
Turner Valley AB 79 F3
Twillingate NL 91 E3
Unity SK 80 B2
Uxbridge ON 84 C3
Val-des-Monts QC 85 E1
Val-d'Or QC 86 A2
Valleyview AB 79 E1
Vancouver BC 78 C4
Vanderhoof BC 78 C1
Vaudreuil-Dorion QC 85 F1
Vaughan ON 84 C3
Vegreville AB 79 F1
Vermilion AB 80 A2
Vernon BC 79 D3
Victoria BC 78 C4
Victoria NL 91 F4
Victoriaville QC 87 D3
Ville-Marie QC 83 F4
Virden MB 81 D4
Vulcan AB 79 F3
Wabana NL 91 F4
Wabasca AB 76 B3
Wabush NL 90 A1
Wadena SK 80 C3
Wainwright AB 80 A2
Wakaw SK 80 C2
Waldheim SK 80 C2
Walkerton ON 84 B3
Wallaceburg ON 84 A4
Warman SK 80 C3
Warwick QC 87 D4
Wasaga Beach ON 84 C2
Waskaganish QC 83 F1
Waterloo ON 84 B3
Waterloo QC 87 D4
Watrous SK 80 C3
Watson Lake YT 75 E3
Wawa ON 83 D4
Wedgeport NS 88 B4
Welland ON 84 C4
Westbank BC 79 D4
Western Shore NS 88 C4
Westlock AB 79 F1
W. Nipissing ON 84 C1
W. Vancouver BC 125 A1
Westville NS 89 D3
Wetaskiwin AB 79 F2
Weyburn SK 80 C4
Whistler BC 78 C3
Whitby ON 84 C3
Whitchurch-Stouffville ON 84 C3
White City SK 80 C4
Whitecourt AB 79 E1
Whitehorse YT 75 D3
Whitewood SK 81 D4
Wilkie SK 80 B2
Williams Lake BC 78 C2
Windsor NS 88 C3
Windsor ON 84 A4
Windsor QC 87 D4
Winkler MB 81 E4
Winnipeg MB 81 F4
Winnipeg Beach MB 81 F3
Witless Bay NL 91 F4
Wolfville NS 88 C3
Wood Buffalo AB 76 B3
Woodstock NB 88 A2
Woodstock ON 84 B3
Wynyard SK 80 C3
Yarmouth NS 88 B4
Yellowknife NT 76 B1
Yorkton SK 81 D3

Vancouver BC

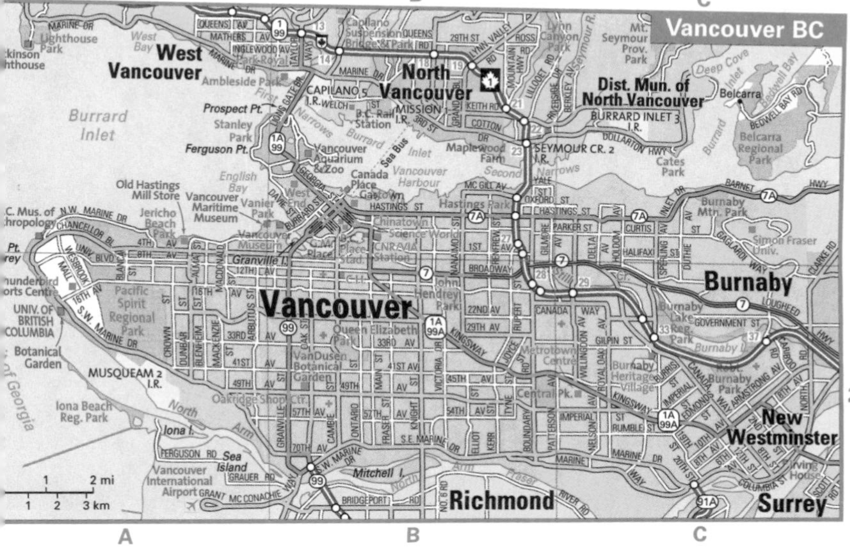

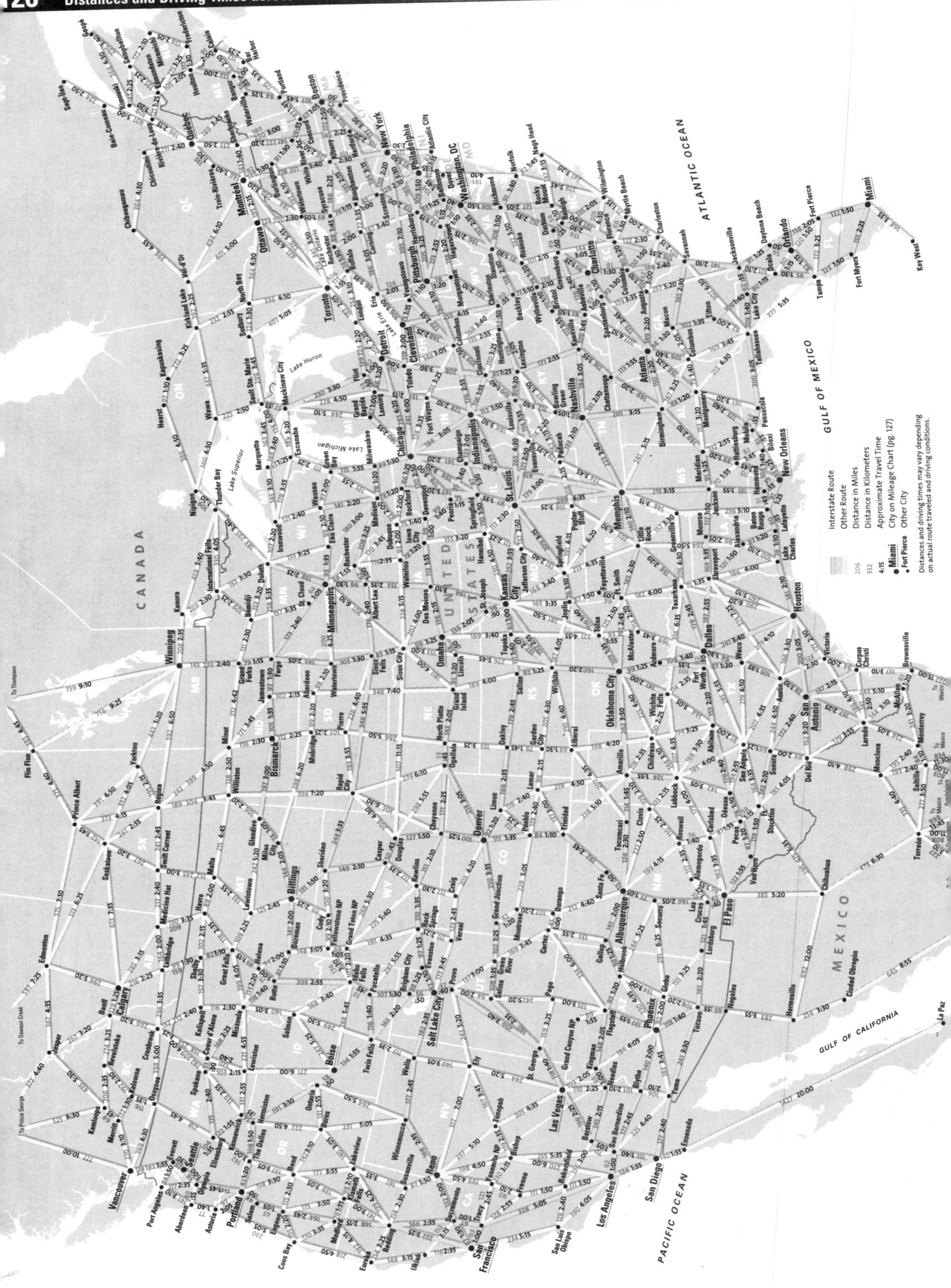
CANADA
UNITED STATES
MEXICO
ATLANTIC OCEAN
PACIFIC OCEAN
GULF OF MEXICO
GULF OF CALIFORNIA
Interstate Route
Other Route
206 Distance in Miles
332 Distance in Kilometers
4:15 Approximate Travel Time
Miami City on Mileage Chart (pg. 127)
Fort Pierce Other City
Distances and driving times may vary depending on actual route traveled and driving conditions.

Miles

Kilometers

City to City Distance Chart (diagonal matrix). City labels along the diagonal:

Albuquerque, NM · Anchorage, AK · Atlanta, GA · Billings, MT · Bismarck, ND · Boise, ID · Boston, MA · Calgary, AB · Charlotte, NC · Chicago, IL · Cleveland, OH · Dallas, TX · Denver, CO · Detroit, MI · El Paso, TX · Halifax, NS · Houston, TX · Indianapolis, IN · Kansas City, MO · Las Vegas, NV · Los Angeles, CA · Memphis, TN · México, MX · Miami, FL · Minneapolis, MN · Montréal, QC · Nashville, TN · New Orleans, LA · New York, NY · Oklahoma City, OK · Omaha, NE · Orlando, FL · Ottawa, ON · Philadelphia, PA · Phoenix, AZ · Pittsburgh, PA · Portland, OR · Québec, QC · Reno, NV · St. Louis, MO · Salt Lake City, UT · San Antonio, TX · San Diego, CA · San Francisco, CA · Seattle, WA · Toronto, ON · Vancouver, BC · Washington, DC · Winnipeg, MB

Tire Tips: Rotation

Regular rotation extends the life of your tires, saving you time and money in the long run. For rotation, each tire and wheel is removed from your vehicle and moved to a different position. This ensures that all of the tires wear evenly and last longer. If no period is specified in your owner's manual, the tires should be rotated every 6–8,000 miles.

Tire Tips: Air Pressure

Keeping your tires properly inflated is essential. We recommend checking air pressure once a month, and before a long trip. Always inflate your tires to the recommended pressure listed by your vehicle's manufacturer. **This information can be found in the owner's manual and often on a placard located in the vehicle's door jamb, inside the fuel hatch, or on the glove compartment door.**

Tire Tips: Alignment

Alignment generally refers to the adjustment of a vehicle's front and rear suspension parts. Proper alignment ensures that your vehicle handles correctly and will help increase the life and performance of your tires. The alignment of your vehicle can be knocked out of adjustment from daily impacts such as potholes and railroad crossings or by more severe accidents. You should have the alignment checked if:
- You know you have hit something.
- You see a wear pattern developing on the shoulder of the tires.
- You notice a difference in your vehicle's handling.

Tire Tips: Good Tread = Good Traction

Worn tread can decrease traction and lead to loss of control. So it's especially important to monitor tread wear ...t and slippery road conditions. Two easy ways to tell if your tires are in need of replacement:

1) Examine the wear bars (narrow bands that appear as grooves across the tread). If they are even with the tread, replace the tire.
2) Place a penny with Lincoln's head down into the shallowest groove of the tire. If the top of his head is visible, then the tire should be replaced.

New Tread

Worn Tread
Tire in need of replacement.

Tire Tips: A Quick Check to Keep You Rolling

1) Determine whether any objects, such as nails or screws, have penetrated the tread.
2) Scan for damage in the tread and sidewall. Look for cuts, cracks, bulges or signs of aging.
3) Check for uneven tread wear. Common causes include:

Over Inflation	**Under Inflation**	**Poor Alignment**	**Out of Balance**
Evident by wear in the middle of the tread.	Look for wear on both shoulders.	Indicated by wear on the inside or outside edge.	Results in wear left or right of center ("cupping").

UNITED STATES
Michelin Travel Publications
One Parkway South
Greenville, South Carolina 29615
TEL 1-800-423-0485, FAX 864-458-5665
EMAIL michelin.travel-publications-us@us.michelin.com

CANADA
Michelin Travel Publications
2540, Boul. Daniel-Johnson, Suite 510
Laval, Québec, H7T 2T9
TEL. 1-800-361-8236, FAX 1-800-361-6937
EMAIL michelin.travel-publications-canada@ca.michelin.com

"Developed using **MAPQUEST** data"

"The information contained herein is for informational purposes only. No representation is made or warranty given as to its content, road conditions or route usability or expeditiousness. User assumes all risk of use. MapQuest, its parents and affiliates, and Michelin North America, Inc. and their respective suppliers assume no responsibility for any loss or delay resulting from such use."